SCIENCE
A CLOSER LOOK

BUILDING SKILLS

Reading and Writing

McGraw Hill Education

Instructions for Copying

Answers are printed in non-reproducible blue. Copy pages on a light setting in order to make multiple copies for classroom use.

Contents

LIFE SCIENCE

Unit Literature 1

Chapter 1 A Look at Living Things
Chapter Concept Map. 2
Lesson 1 Lesson Outline 3
Lesson Vocabulary 5
Lesson Cloze Activity 6
Reading in Science 7
Lesson 2 Lesson Outline 9
Lesson Vocabulary 11
Lesson Cloze Activity 12
Lesson 3 Lesson Outline 13
Lesson Vocabulary 15
Lesson Cloze Activity 16
Lesson 4 Lesson Outline 17
Lesson Vocabulary 19
Lesson Cloze Activity 20
Writing in Science 21
Chapter Vocabulary 23

Chapter 2 Living Things Grow and Change
Chapter Concept Map. 25
Lesson 1 Lesson Outline 26
Lesson Vocabulary 28
Lesson Cloze Activity 29
Lesson 2 Lesson Outline 30
Lesson Vocabulary 32
Lesson Cloze Activity 33
Writing in Science 34
Lesson 3 Lesson Outline 36
Lesson Vocabulary 38
Lesson Cloze Activity 39
Reading in Science 40
Chapter Vocabulary 42

Contents

Unit Literature 44

Chapter 3 Living Things in Ecosystems

Chapter Concept Map 45

Lesson 1 Lesson Outline 46
                Lesson Vocabulary 48
                Lesson Cloze Activity 49

Lesson 2 Lesson Outline 50
                Lesson Vocabulary 52
                Lesson Cloze Activity 53
                Reading in Science 54

Lesson 3 Lesson Outline 56
                Lesson Vocabulary 58
                Lesson Cloze Activity 59

Chapter Vocabulary 60

Chapter 4 Changes in Ecosystems

Chapter Concept Map 62

Lesson 1 Lesson Outline 63
                Lesson Vocabulary 65
                Lesson Cloze Activity 66

Lesson 2 Lesson Outline 67
                Lesson Vocabulary 69
                Lesson Cloze Activity 70
                Writing in Science 71

Lesson 3 Lesson Outline 73
                Lesson Vocabulary 75
                Lesson Cloze Activity 76
                Reading in Science 77

Chapter Vocabulary 79

Contents

EARTH SCIENCE

Unit Literature 81

Chapter 5 Earth Changes

Chapter Concept Map. 82

Lesson 1 Lesson Outline 83
Lesson Vocabulary 85
Lesson Cloze Activity 86

Lesson 2 Lesson Outline 87
Lesson Vocabulary 89
Lesson Cloze Activity 90
Reading in Science 91

Lesson 3 Lesson Outline 93
Lesson Vocabulary 95
Lesson Cloze Activity 96
Writing in Science 97

Chapter Vocabulary 99

Chapter 6 Using Earth's Resources

Chapter Concept Map. 101

Lesson 1 Lesson Outline 102
Lesson Vocabulary 104
Lesson Cloze Activity 105
Writing in Science 106

Lesson 2 Lesson Outline 108
Lesson Vocabulary 110
Lesson Cloze Activity 111

Lesson 3 Lesson Outline 112
Lesson Vocabulary 114
Lesson Cloze Activity 115
Reading in Science 116

Lesson 4 Lesson Outline 118
Lesson Vocabulary 120
Lesson Cloze Activity 121

Chapter Vocabulary 122

Contents

Unit Literature 124

Chapter 7 Changes in Weather

Chapter Concept Map 125

Lesson 1 Lesson Outline 126
          Lesson Vocabulary 128
          Lesson Cloze Activity 129

Lesson 2 Lesson Outline 130
          Lesson Vocabulary 132
          Lesson Cloze Activity 133
          Reading in Science 134

Lesson 3 Lesson Outline 136
          Lesson Vocabulary 138
          Lesson Cloze Activity 139
          Writing in Science 140

Chapter Vocabulary 142

Chapter 8 Planets, Moons, and Stars

Chapter Concept Map 144

Lesson 1 Lesson Outline 145
          Lesson Vocabulary 147
          Lesson Cloze Activity 148
          Writing in Science 149

Lesson 2 Lesson Outline 151
          Lesson Vocabulary 153
          Lesson Cloze Activity 154

Lesson 3 Lesson Outline 155
          Lesson Vocabulary 157
          Lesson Cloze Activity 158

Lesson 4 Lesson Outline 159
          Lesson Vocabulary 161
          Lesson Cloze Activity 162
          Reading in Science 163

Chapter Vocabulary 165

Contents

PHYSICAL SCIENCE

Unit Literature 167

Chapter 9 Observing Matter

Chapter Concept Map. 168

Lesson 1 Lesson Outline 169

Lesson Vocabulary 171

Lesson Cloze Activity 172

Reading in Science 173

Lesson 2 Lesson Outline 175

Lesson Vocabulary 177

Lesson Cloze Activity 178

Lesson 3 Lesson Outline 179

Lesson Vocabulary 181

Lesson Cloze Activity 182

Writing in Science 183

Chapter Vocabulary 185

Chapter 10 Changes in Matter

Chapter Concept Map. 187

Lesson 1 Lesson Outline 188

Lesson Vocabulary 190

Lesson Cloze Activity 191

Lesson 2 Lesson Outline 192

Lesson Vocabulary 194

Lesson Cloze Activity 195

Reading in Science 196

Lesson 3 Lesson Outline 198

Lesson Vocabulary 200

Lesson Cloze Activity 201

Chapter Vocabulary 202

Contents

Unit Literature 204

Chapter 11 Forces and Motion

Chapter Concept Map. 205

Lesson 1 Lesson Outline 206

Lesson Vocabulary 208

Lesson Cloze Activity 209

Reading in Science 210

Lesson 2 Lesson Outline 212

Lesson Vocabulary 214

Lesson Cloze Activity 215

Lesson 3 Lesson Outline 216

Lesson Vocabulary 218

Lesson Cloze Activity 219

Lesson 4 Lesson Outline 220

Lesson Vocabulary 222

Lesson Cloze Activity 223

Writing in Science 224

Chapter Vocabulary 226

Chapter 12 Forms of Energy

Chapter Concept Map. 228

Lesson 1 Lesson Outline 229

Lesson Vocabulary 231

Lesson Cloze Activity 232

Lesson 2 Lesson Outline 233

Lesson Vocabulary 235

Lesson Cloze Activity 236

Lesson 3 Lesson Outline 237

Lesson Vocabulary 239

Lesson Cloze Activity 240

Reading in Science 241

Lesson 4 Lesson Outline 243

Lesson Vocabulary 245

Lesson Cloze Activity 246

Writing in Science 247

Chapter Vocabulary 249

Monarch
by Marilyn Singer

Read the Unit Literature feature in your textbook.

Write About It

Response to Literature This poem describes a caterpillar changing into a butterfly. All living things change as they grow. Write a poem about how you have changed as you have grown. Write about some exciting things you are waiting for.

Students' poems should demonstrate vivid word choice and include

grammar and mechanics. Their poems should include students' own

observations about the changes in their lives and hopes for the

future.

_____

_____

_____

_____

_____

_____

_____

© Macmillan/McGraw-Hill

Name _____ Date _____

A Look at Living Things

Complete the concept map about structures of plants and animals. Some examples have been done for you.

What Living Things Need	Plant Structures	Animal Structures
Food	▶ Sugars are made inside _leaves_ during photosynthesis. ▶ Nutrients are taken in by _roots_.	▶ Food is taken in by _tongues_, _beaks_, _trunks_, and teeth.
Water	▶ Water is taken in by _roots_. ▶ Water flows through _stems_.	▶ Water is taken in by _tongues_, _beaks_, and _trunks_.
Gases	▶ Carbon dioxide is taken in by _leaves_.	▶ Oxygen is taken in by _lungs_, _gills_, and skin.

Living Things and Their Needs

Use your textbook to help you fill in the blanks.

What are living things?

1. A sunflower plant changes with age, or _____grows_____ .

2. A plant responds to shade when it bends toward

   _____sunlight_____ .

3. Trees reproduce by making _____seeds_____ .

4. Alligators lay _____eggs_____ to make more of
   their own kind.

5. Water and sunlight are _____nonliving_____ things
   in nature.

6. Rocks are nonliving because they do not grow,

   respond, or _____reproduce_____ .

What do living things need?

7. All living things need food, water, space, and

   _____gases_____ in order to survive.

8. Animals eat other organisms because they need

   _____food_____ for energy.

9. Living things need _____water_____ to break
   down food.

10. Air and water contain a(n) _____gas_____
    called oxygen.

11. Plants need the gases oxygen and _____carbon dioxide_____ to survive.

12. Living things need room, or _____space_____ , to grow, move, and find food.

13. Living and nonliving things are part of an organism's

_____environment_____ .

What are living things made of?

14. Living things are made of small parts called _____cells_____ .

15. A tool called a(n) _____microscope_____ helps us to see cells.

16. Organisms that have one cell and live in many places

are called _____bacteria_____ .

Critical Thinking

17. What characteristics can **not** be used to tell the difference between living and nonliving things?

Possible answers: size, shape, color, movement, sound, the

environment in which something is located, temperature,

complex structure

_____

_____

_____

Living Things and Their Needs

Match the correct letter with the description.

---

a. carbon dioxide	**d.** microscope	**g.** reproduce
b. cell	**e.** organism	**h.** respond
c. environment	**f.** oxygen	

---

1. ____g____ to make more of one's own kind

2. ____b____ a small part that makes up all living things

3. ____c____ all of the living and nonliving things that surround an organism

4. ____f____ a gas that plants and animals need

5. ____d____ a special tool that helps make tiny things look larger

6. ____a____ a gas that plants use to make food

7. ____e____ another name for a living thing

8. ____h____ to react to the world around you

Living Things and Their Needs

Fill in the blanks. Use the words from the box.

carbon dioxide	food	oxygen
cells	grow	reproduce
energy	organisms	respond

Living things are made of small parts called cells.

Some organisms are made of many _____cells_____ .

Others are made of only one cell.

Living things have needs. They need food for

_____energy_____ to help them move and _____grow_____ .

They need water to break down and move _____food_____

through their bodies. They need gases. Animals get the

gas _____oxygen_____ from air or water. Plants also need

the gas _____carbon dioxide_____ .

Living things, or _____organisms_____ , have many

characteristics in common. They _____respond_____ when

they are in danger or when they get too hot. Living things

_____reproduce_____ to make new plants and animals. A

thing without these characteristics is nonliving .

© Macmillan/McGraw-Hill

Eating Away at Pollution

Read the Reading in Science feature in your textbook.

Write About It

Classify The article explains that some microorganisms are harmful and others are helpful. This is a way to classify them. Read the article again with a partner. Look for another way to classify microorganisms. Then write about it.

Classify

Fill in the blanks in the graphic organizer below. When you have finished, you will be able to see how the two groups are alike and different.

	Harmful Microorganisms	Helpful Microorganisms
Where they are	They are ___all around___ us.	They are ___all around___ us.
What they do	They make plants and ___animals___ sick.	They get rid of ___pollution___ .
Size	They are ___tiny___ .	They are ___tiny___ .

Planning and Organizing

Answer the following questions.

What do helpful microorganisms eat?

Some eat things that are harmful to animals and plants, such as oil

and dangerous chemicals found in smoke.

What do helpful microorganisms help clean?

Some clean water and land. Others help clean the air.

_____

Drafting

Explain how helpful microorganisms are alike.

They eat pollution and keep Earth clean. They are small living things.

_____

Explain how helpful microorganisms are different.

Some eat oil, and others eat chemicals. Some clean the water and

soil. Others clean the air.

Now, write how you would classify helpful microorganisms.

I would classify them by whether or not they eat oil or chemicals.

_____

Plants and Their Parts

Use your textbook to help you fill in the blanks.

What are plants?

1. One way in which plants are alike is that they make

   their own _____ food _____ .

2. Roots, stems, and leaves help a plant get what it

   needs in order to _____ survive _____ .

How do roots and stems help plants?

3. Roots collect water from the _____ soil _____ .

4. Carrots have a thick root, or _____ taproot _____ , that
   stores food.

5. Roots take in _____ nutrients _____ that help plants grow
   and stay healthy.

6. The plant part that holds up leaves to sunlight is the

   _____ stem _____ .

7. Stems have _____ tubes _____ that carry food and
   water through a plant.

8. A tree has a hard, woody stem called

   a(n) _____ trunk _____ .

Why are leaves important?

9. Leaves are important because _____food_____ is made in the leaves.

10. Plants get energy from the _____Sun_____ to make food.

11. Plants change carbon dioxide and water into _____sugars_____ that they use for food.

12. Leaves have tiny holes that take in _____carbon dioxide_____ .

13. When you breathe, you take in _____oxygen_____ that plants release.

How can you classify plants?

14. Scientists study plants by putting them into _____groups_____ .

15. Scientists group plants by their _____structures_____ , such as roots, stems, or leaves.

Critical Thinking

16. What are the jobs of roots, stems, and leaves?

   Roots collect water, take in nutrients, and hold a plant in place.

   Stems support leaves and have hollow tubes through which

   water, nutrients, and food move. Leaves take in carbon dioxide

   and sunlight and are the places where food is made.

Plants and Their Parts

Use the words in the box to fill in the blank.

leaf	photosynthesis	stem	sugar
nutrient	root	structure	tubes

1. A _____stem_____ is a structure that holds up a plant.

2. A plant's _____structure_____ helps it survive.

3. The structure in which food is made is the ___leaf___ .

4. The process of making food is called ___photosynthesis___ .

5. A _____root_____ is a structure that takes in water.

6. A substance that helps living things grow and stay healthy is a _____nutrient_____ .

7. The substance that is made in the leaves and used as food for the plant is _____sugar_____ .

8. Stems use _____tubes_____ to carry water and food throughout the plant.

Name _____ Date _____

Plants and Their Parts

Fill in the blanks. Use the words from the box.

carbon dioxide	oxygen	structures
classify	photosynthesis	sunlight
nutrients	reproduce	water

All plants have one thing in common. They can make

their own food through the process of _photosynthesis_ .

In this process, plants make sugars from _carbon dioxide_

and water. Plants give off _oxygen_ , which

animals need in order to live.

Most plants also have parts, or _structures_ ,

in common. Scientists use these to group, or _classify_ ,

plants. Flowers and cones help some plants _reproduce_ .

Roots hold plants in place and take in _water_

and nutrients. Water, food, and _nutrients_ flow

through the tubes in stems. Stems help leaves to get

sunlight . Inside a leaf, a plant makes food.

Animals and Their Parts

Use your textbook to help you fill in the blanks.

What are animals?

1. One characteristic that most animals have in common

   is that they can _____ move _____ .

2. Unlike plants, animals cannot make their

   own _____ food _____ .

3. Animals are able to _____ respond _____ to their
   environments much more noticeably than plants.

4. Wings and tails are _____ structures _____ that help
   animals get what they need.

5. Animals move toward food and away

   from _____ danger _____ .

6. Wolves can run and jump because of their

   _____ strong legs _____ .

7. Fins and tails help _____ fish _____ move through
   the water.

8. Ducks and geese have _____ webbed feet _____ that help
   them swim.

How do animals get what they need?

9. Birds use structures called _____ beaks _____ to get
   food and water.

10. Lions use their sharp _____ teeth _____ for biting.

11. Animals that have flat back teeth use them

for _____ chewing _____ .

12. Lungs and gills help animals take in _____ oxygen _____ .

How do animals stay safe?

13. A _____ shelter _____ helps protect animals from bad weather.

14. Rocks and _____ trees _____ are examples of safe places, or shelters, for animals.

15. A snail has a(n) _____ hard shell _____ to keep it safe in its environment.

Critical Thinking

16. A sponge is an animal. Adult sponges do not move. Why do you think a sponge is classified as an animal?

A sponge cannot make its own food. It can take in oxygen from

its environment.

_____

_____

_____

_____

Animals and Their Parts

Match the correct letter with the description.

a. gills	**c.** muscles	**e.** quills	**g.** trunk
b. lung	**d.** nest	**f.** shelter	**h.** wings

1. ____d____ a kind of shelter for young birds

2. ____b____ a structure that takes in oxygen from air

3. ____a____ structures that take in oxygen from water

4. ____g____ a structure that helps elephants pull food to their mouths

5. ____h____ structures that help birds fly and glide through the air

6. ____f____ a safe place for animals

7. ____e____ structures that help protect a porcupine from other animals

8. ____c____ structures that help a snake move

Animals and Their Parts

Fill in the blanks.

air	move	respond	teeth
fins	organisms	swim	

Animals share certain characteristics that make

them different from plants. Animals can fly, run, jump, or

_____ swim _____ . They eat other _____ organisms _____

instead of making their own food. Animals _____ respond _____

to their environments more noticeably than plants.

Animals have a variety of structures. They have

different kinds of _____ teeth _____ for biting and

chewing. Lungs and gills help animals get oxygen from

_____ air _____ and water. Feet, legs, and wings help

different animals _____ move _____ . Fish have tails and

_____ fins _____ that help them move through water.

Tongues, beaks, and trunks help animals get water

and food.

© Macmillan/McGraw-Hill

Classifying Animals

Use your textbook to help you fill in the blanks.

How can you classify animals?

1. Scientists group, or _____classify_____ , animals so that they can study them.

2. One way to classify animals is by the presence or absence of a _____backbone_____ .

What are some invertebrates?

3. The covering around an insect is thin and _____hard_____ .

4. A spider is an invertebrate because it is covered by a(n) _____exoskeleton_____ .

5. Crabs have an exoskeleton called a _____shell_____ .

What are some vertebrates?

6. Birds have lightweight bones and wings to help them _____fly_____ .

7. Birds and _____reptiles_____ have lungs and they lay eggs.

8. Reptiles live in wet or dry places because their skin is _____waterproof_____ .

9. Frogs and toads breathe through _____gills_____ when they are young.

10. Amphibians can live on land because they grow lungs

and _____legs_____ .

11. Fish have scales and a _____flat_____ shape to
move easily through water.

What are mammals?

12. People are a kind of vertebrate called

a _____mammal_____ .

13. Mammals feed _____milk_____ to their young.

14. Some mammals are covered with

thick _____fur_____ .

15. Mammals use their _____lungs_____ to breathe.

Critical Thinking

16. Why are vertebrates classified into different groups?
Give examples of structures for each group.

Vertebrates are classified into different groups because they

have different structures. A bird has a beak, feathers, and wings.

Reptiles have scaly skin and lungs. Amphibians have gills when

they are young, but they grow lungs and legs as they get older.

Fish have scales and gills. Mammals have hair or fur.

_____

© Macmillan/McGraw-Hill

Classifying Animals

What am I?

Choose a word from the word box below to answer each question.

amphibian	fish	mammal
exoskeleton	invertebrate	vertebrate

1. I am a hard, outer covering around animals with no backbone. What am I? _____exoskeleton_____

2. I am a vertebrate that spends part of my life in water and part of my life on land. What am I? _____amphibian_____

3. I have a soft body and no backbone. What am I? _____invertebrate_____

4. I am an animal with a backbone. What am I? _____vertebrate_____

5. I am a vertebrate with gills and I live in water my whole life. What am I? _____fish_____

6. I am a vertebrate with hair. I feed milk to my young and care for them. What am I? _____mammal_____

Name _____ Date _____

Classifying Animals

Use the words in the box to fill in the blanks.

amphibian	exoskeleton	invertebrates
backbone	fish	reptile
bird	gills	vertebrates

Scientists classify animals into groups to make

studying them easier. Two main groups of animals are

vertebrates and _____invertebrates_____ . Invertebrates have

an _____exoskeleton_____ to protect their bodies. There are

many more invertebrates than vertebrates.

There are five kinds of _____vertebrates_____ . All of

them have a _____backbone_____ to support their bodies.

The kind of vertebrate that lays eggs and can fly is a

_____bird_____ . A _____reptile_____ has scaly skin

and breathes with lungs. An _____amphibian_____ looks

like a fish when it hatches. Its _____gills_____ change

to lungs as the animal gets older. Vertebrates that

swim and have scales are called _____fish_____ .

Mammals, another kind of vertebrate, do not hatch

from eggs but are born.

Use with **Lesson 4**
Classifying Animals

© Macmillan/McGraw-Hill

Desert Birds

Read the Writing in Science feature in your textbook.

Write About It

Descriptive Writing Choose two animals. Learn more about them. Then write a paragraph on a separate piece of paper that describes how the animals are alike and different.

Getting Ideas

Select two different animals. Write the name of each animal above each oval below. In the outer part of each oval, write how each animal is different. In the part that overlaps, tell how they are the same.
Possible answer:

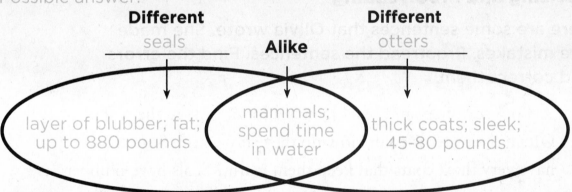

Different
seals

Alike

Different
otters

layer of blubber; fat; up to 880 pounds

mammals; spend time in water

thick coats; sleek; 45-80 pounds

Planning and Organizing

Olivia wanted to show how a sea otter and a seal are alike and different. Here are two sentences that she wrote. Write "compare" if the sentence tells how they are alike, and "contrast" if it tells how they are different.

1. _____compare_____ Both seals and sea otters are mammals that spend time in the water.

2. _____contrast_____ Unlike seals, sea otters do not have a layer of blubber.

Drafting

Begin your description by writing your own sentence. Name your two animals. Write a main idea that includes them.

Possible answer: Both sea otters and seals spend a lot of time in the water, but they are otherwise very different.

Now write a paragraph describing your two animals. Use a separate piece of paper. Write how they are alike and different. Include vivid details to describe the animals.

Revising and Proofreading

Here are some sentences that Olivia wrote. She made five mistakes. Proofread the sentences. Find the errors and correct them.

Both sea otters and seals spends a lot of time in the water. Often, the water is cold. How do they stay warm. Sea otters have very thick coats that keep them warm? Seals have blubber to protekt them from the cold. Usually, sea otters look sleek, while seals looks fat.

Now revise and proofread your writing. Ask yourself:

▶ Did I show how my two animals are alike and different?

▶ Did I include descriptive words?

▶ Did I correct all mistakes?

A Look at Living Things

Circle the letter of the best answer.

1. A living thing that makes more of its own kind is said to

   a. grow.

   b. respond.

   c. reproduce.

   d. react.

2. The living and nonliving things that surround an organism make up its

   a. shelter.

   b. environment.

   c. exoskeleton.

   d. structure.

3. The structure in which a plant makes food is a

   a. stem.

   b. leaf.

   c. root.

   d. flower.

4. The plant structures that collect water and nutrients are

   a. roots.

   b. stems.

   c. leaves.

   d. flowers.

5. Every living thing can be called a(n)

   a. cell.

   b. invertebrate.

   c. vertebrate.

   d. organism.

Circle the letter of the best answer.

6. A structure that land animals use to breathe air is a(n)

    a. lung.

    b. gill.

    c. exoskeleton.

    d. cell.

7. An animal without a backbone is a(n)

    a. fish.

    b. amphibian.

    c. vertebrate.

    d. invertebrate.

8. An animal that has feathers and breathes with lungs is a(n)

    a. amphibian.

    b. reptile.

    c. bird.

    d. mammal.

9. An animal that has gills and lungs during its life is a(n)

    a. fish.

    b. reptile.

    c. amphibian.

    d. invertebrate.

10. A vertebrate that has hair and gives birth to its young is a(n)

    a. amphibian.

    b. reptile.

    c. bird.

    d. mammal.

Living Things Grow and Change

Complete the chart below to show the stages in the life cycles of plants and animals. Some answers have been completed for you.

Flowering Plants

| Seeds made in flowers | → | seedling | → | adult | → | plant dies |

Conifers

| Seeds made in cones | → | seedling | → | adult | → | plant dies |

Amphibians and Most Insects

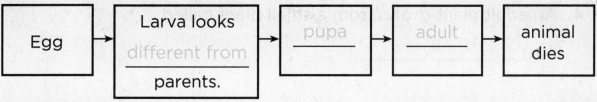

| Egg | → | Larva looks different from parents. | → | pupa | → | adult | → | animal dies |

Reptiles and Fish

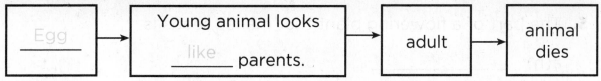

| Egg | → | Young animal looks like parents. | → | adult | → | animal dies |

Birds

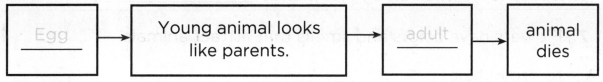

| Egg | → | Young animal looks like parents. | → | adult | → | animal dies |

Mammals

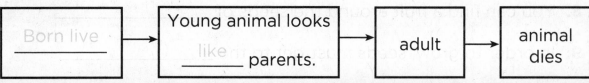

| Born live | → | Young animal looks like parents. | → | adult | → | animal dies |

Name _____ Date _____

Plant Life Cycles

Use your textbook to help you fill in the blanks.

How do plants grow?

1. The structure inside an apple that grows into a new

   plant is a(n) _____seed_____ .

2. A seed has stored food and nutrients to help the

   _____embryo_____ survive.

3. When conditions are right, a seed will begin to grow,

   or _____germinate_____ .

4. An adult plant grows from a small plant called

   a(n) _____seedling_____ .

How do plants make seeds?

5. The part of a flowering plant that makes seeds is

   a(n) _____flower_____ .

6. Seeds form when an egg joins with _____pollen_____ .

7. Flowers have colors and smells that attract animals

   to drink their _____nectar_____ .

8. You can find a fruit around the seeds of _____flowering plants_____ .

9. In order to grow, seeds must get to the _____soil_____ .

What is the life cycle of some plants?

10. A seed germinates in the first stage of a flowering

plant's _____life cycle_____ .

11. When plants die, they add _____nutrients_____ to
the soil.

12. Two kinds of plants that reproduce by making seeds

are flowering plants and _____conifers_____ .

13. Pollen moves from small male cones to large female

cones when the _____wind_____ blows.

How do plants grow without seeds?

14. An onion can grow a new plant from its underground

stem, or _____bulb_____ .

Critical Thinking

15. What are the steps in the life cycle of a flowering
plant? Use the terms *seed*, *germinate*, *seedling*,
flower, and *pollination* in your answer.

Flowering plants start as seeds, which germinate and grow into

seedlings. The seedlings grow into adult plants, which produce

flowers that make seeds through the process of pollination.

Then a seed develops, and the cycle starts again. When adult

plants die, they break down, adding nutrients to the soil that

new plants use to grow.

Plant Life Cycles

Choose a word from the box that matches each clue below and write its letter in the space provided.

a. cone	**c.** flower	**e.** life cycle	**g.** pollination
b. embryo	**d.** fruit	**f.** pollen	**h.** seed

1. ___h___ a structure that can grow into a new plant

2. ___c___ a structure in flowering plants that makes seeds

3. ___e___ all of the stages in an organism's life

4. ___g___ the process that takes place when pollen moves from the male part of a flower to the female part of a flower

5. ___d___ a structure that holds seeds

6. ___b___ a young plant inside a seed

7. ___a___ a structure in conifers that makes seeds

8. ___f___ a powder made by the male part of a flower or cone

© Macmillan/McGraw-Hill

Plant Life Cycles

Use the words in the box to fill in the blanks below.

adult	fruit	reproduce
cones	germinate	wind
eggs	pollination	

Plants go through stages known as a life cycle.

Plants _____germinate_____ from seeds and grow into

_____adult_____ plants. Then the plants reproduce.

When plants die, they return nutrients to the soil that

new plants use.

Flowers help flowering plants _____reproduce_____ .

Flowers produce pollen and _____eggs_____ .

Animals and _____wind_____ move pollen to eggs.

This movement is called _____pollination_____ . After a

flower is pollinated, a seed forms and is protected by a

_____fruit_____ that grows around it. Conifers make

seeds in _____cones_____ instead of flowers. Wind

blows pollen from small male cones to large female

cones. The large cones grow seeds.

Animal Life Cycles

Use your textbook to help you fill in the blanks.

What are some animal life cycles?

1. A _____tadpole_____ changes into a frog as it grows.

2. Animals change in different ways, but all change as

   part of their _____life cycles_____ .

3. After an animal is born, it grows, changes, _____reproduces_____ ,
   and dies.

4. During their life cycles, some animals change form

   through the process of _____metamorphosis_____ .

5. Metamorphosis happens in the life cycles of

   amphibians and some _____insects_____ .

6. The life cycle of amphibians and insects begins with

   a(n) _____egg_____ .

7. A young amphibian that _____hatches_____ from an
   egg does not look like an adult.

8. Another name for an insect that has just hatched

   is _____larva_____ .

**How do reptiles, fish, and birds change
as they grow?**

9. Fish lay their eggs in _____water_____ .

© Macmillan/McGraw-Hill

10. When reptiles and fish are young, they look like
_____adults_____ , similar to their parents.

11. Unlike most reptiles and fish, _____birds_____
protect their eggs and raise their young.

What is the life cycle of a mammal?

12. Most mammals do not hatch from eggs, but are
_____born live_____ .

13. Like birds, young mammals _____look_____
like adults.

14. Mammals look after their young until the young can
_____survive_____ on their own.

Critical Thinking

15. How are the life cycles of animals alike and different?

The life cycles of animals are alike because they are born or

hatched. They all grow and change, reproduce, and die. They

are different because most animals hatch from eggs, but

mammals are born live. Only amphibians and some insects

go through metamorphosis. Unlike most animals, birds and

mammals take care of their young until the young can live on

their own.

Animal Life Cycles

Match the correct letter with its description.

a. adult	**d.** larva	**g.** pupa
b. egg	**e.** life cycle	**h.** tadpole
c. hatching	**f.** metamorphosis	

1. ___b___ a structure containing food and nutrients that young animals need in order to grow

2. ___e___ the stages through which animals grow, change, reproduce, and die

3. ___c___ the process by which an animal breaks out of an egg

4. ___d___ a young insect that has just hatched

5. ___a___ the stage of an animal's life cycle when it reproduces

6. ___g___ the stage in which an insect is changing into an adult

7. ___f___ a process by which an organism's body changes form

8. ___h___ a young frog that breathes with gills

Animal Life Cycles

Use the words in the box to fill in the blanks below.

adults	larva	live	parents
die	lay eggs	mammals	
hatch	life cycles	metamorphosis	

Animals grow, change, and reproduce in different

ways. All animals change during their _____life cycles_____ .

Animals are hatched from eggs or born _____live_____ .

At the end of their life cycles, all animals _____die_____ .

Reptiles, fish, and birds all _____lay eggs_____ and

_____mammals_____ are born live. Young birds and

mammals look similar to their _____parents_____ . When

they _____hatch_____ , young reptiles and fish look

just like their parents. Amphibians and insects in the

_____larva_____ stage look very different from their

parents. Larvae hatch from eggs, and then change into

_____adults_____ through a process called

_____metamorphosis_____ . They will then look like

their parents.

Name _____ Date _____

The Little Lambs

Read the Writing in Science feature in your textbook.

Write About It

Personal Narrative Have you ever seen a plant or animal grow and change? Write about your experience. Describe the changes. Write what you observed, what you did, and how it made you feel.

Getting Ideas

Select a plant or animal to write about. Think about how it changed as it grew. Write three stages of its growth down in the sequence chart below.

Plant or animal: _____ horse _____

My foal had long legs and no teeth.	→ My horse got its first set of teeth when it was a year old.	→ My horse is still growing and gaining weight at three years old.

Planning and Organizing

Jake wrote about his horse Wind Star. Here are three sentences that he wrote. Put them in time order.
Write 1 next to the sentence that should come first.
Write 2 next to the sentence that should come next.
Write 3 next to the sentence that should come last.

1. ____2____ Wind Star got his first set of teeth when he was one.

2. ____3____ Now he is three years old and still growing.

3. ____1____ When Wind Star was a foal, he had no teeth.

Drafting

Write the first sentence of your narrative. Use "I" to refer to yourself. Describe something interesting about a plant or animal that you helped to care for.

Possible sentence: I fell in love with Wind Star the first time I saw him.

Now complete your personal narrative. Use a separate piece of paper. Begin with the sentence you wrote above. Include details about how your plant or animal grew and changed. Put them in time order. Explain how watching these changes made you feel.

Revising and Proofreading

Here is part of the personal narrative that Jake wrote. He had a lot of trouble with homophones. Homophones are words that sound alike but have different spellings and different meanings. Proofread it. Find the five mistakes he made. Correct them.

My little foul looked so handsome. He had a white star write in the middle of his forehead. His coat was chestnut brown. His legs wobbled whenever he stood up. He was the cutest creature I had ever scene. I couldn't weight for him to grow up sew that I could ride on him.

Now revise and proofread your own writing.
Ask yourself:

▶ Did I use the pronoun "I" to describe my own experience?

▶ Did I detail how the plant or animal grew and changed?

▶ Did I correct all mistakes?

From Parents to Young

Use your textbook to help you fill in the blanks.

What are inherited traits?

1. Features that make an organism unique are

   called _____traits_____ .

2. Examples of human traits are eye and hair _____color_____ .

3. People use traits to _____describe_____ an organism.

4. The passing on of a trait from parents to young is

   called _____heredity_____ .

5. The features that your parents passed on to you are

   called _____inherited traits_____ .

6. You look like your _____parents_____ because of
   inherited traits.

7. Most organisms have a(n) _____mixture_____ of traits
   from both parents.

8. An organism will look _____more_____ like the
   parent that passes on visible traits.

9. If a gray dog and a yellow dog have yellow puppies,

   yellow is a more _____visible_____ trait than gray.

10. Both parents pass on traits to their young, otherwise

    known as _____offspring_____ .

Which traits are not inherited?

11. Traits that people and animals learn over time are

called _____learned traits_____ .

12. Learned traits and traits caused by the environment

are not _____inherited_____ .

13. A scar is a trait caused by the _____environment_____ .

14. An example of a trait caused by the environment is a

tree losing its _____branches_____ .

Critical Thinking

15. What is the difference between traits that are
inherited and traits that are not inherited? Give
examples in your answer.

Traits that are inherited come from parents. Examples are hair

and eye color. Traits that are learned or are caused by the

environment are not inherited. An example of a learned trait

is riding a bicycle. An example of a trait that is caused by the

environment is a rabbit that gets fat when it eats plenty of food.

_____

_____

_____

_____

From Parents to Young

Fill in the blanks.

environment	inherited traits	offspring	visible
heredity	learned traits	trait	

1. Traits that come from parents are called _____ . *inherited traits*

2. You may look more like one of your parents than the other one because that parent passed on _____ traits. *visible*

3. New skills you gain that were not passed on from your parents are called _____ . *learned traits*

4. The passing on of traits from parents to young is called _____ . *heredity*

5. A unique feature of a living thing is a(n) _____ . *trait*

6. Hair that gets lighter from the Sun is an example of a trait caused by the _____ . *environment*

7. Another word for an organism's young is _____ . *offspring*

From Parents to Young

Use the words in the box to fill in the blanks below.

environment	learned traits	offspring
inherited traits	mixture	skills

The features that make an organism unique are

called traits. Some traits are passed from parents to

their ____offspring____ . These traits are called

____inherited traits____ . Organisms look like both parents

because they get a(n) ____mixture____ of traits from

both parents. However, an organism may look more

like the parent that has passed on the more visible

traits.

Other traits are new ____skills____ , such as

learning to read. They are not inherited. Skills that an

animal learns are called ____learned traits____ . Learned

traits can change or appear because of the ____environment____ .

For example, green leaves may turn yellow if a plant

needs water.

© Macmillan/McGraw-Hill

Name _____ Date _____

Meet Darrel Frost

Read the text below.

A fence lizard is soaking up warmth from the Sun when a hawk soars overhead. The hawk spots the lizard, then swoops down to grab a meal. The lizard has no time to scurry away. How can it escape the hawk's claws?

If the hawk catches the lizard's long tail, the tail will break off. The bird will be left holding a wriggling tail while the lizard runs away. In time, the lizard will grow a new tail. Growing new body parts is a trait called regeneration.

Regeneration is one of the many amazing traits that Darrel Frost studies. Darrel is a scientist at the American Museum of Natural History. He travels all over the world to learn about different kinds of lizards. Then he observes their traits. Finally, he uses his observations to find out how different kinds of lizards are related.

Rewrite the first two sentences using the words *first*, *second*, *third*, **and** *fourth*.

First, a fence lizard is soaking up warmth from the Sun. Second, a

hawk soars overhead. Third, the hawk spots the lizard. Fourth, the

hawk swoops down to grab a meal.

_____

_____

_____

Write About It

Sequence Read the article with a partner. Fill in a sequence-of-events chart to show how Darrel learns about lizards. Then use your chart to write a summary about Darrel and his work.

> **First**
> Darrel travels all over the world to learn about lizards.

> **Next**
> He observes them and studies their traits.

> **Last**
> He uses his observations to try to find out how different kinds of lizards are related.

Summarize

Write your summary on the lines below. Use your own words. Include the ideas in the boxes above. Be sure your summary tells what happens first, next, and last.

Darrel is a scientist who studies amphibians and reptiles. He studies regeneration, which means growing a new body part. Darrel is interested in learning about types of lizards. First, Darrel travels around the world. Next, he observes and studies the lizards. Last, he tries to find out how different lizards are related.

Living Things Grow and Change

Circle the letter of the best answer.

1. Which structure of a
flowering plant holds seeds?

   a. cone

   b. fruit

   c. egg

   d. embryo

2. Which structure contains
food and nutrients for
developing animals?

   a. embryo

   b. egg

   c. seed

   d. pupa

3. When pollen joins an egg in
the ovary of a flower, what is
formed?

   a. a seed

   b. a cone

   c. a flower

   d. a fruit

4. Which is an example of a
learned trait?

   a. flower shape

   b. eye color

   c. speaking a language

   d. a scar

5. Which young animal looks
very different from its
parents?

   a. kitten

   b. alligator

   c. chick

   d. beetle larva

6. Which plant structure makes
seeds?

   a. cone

   b. fruit

   c. flower

   d. embryo

Circle the letter of the best answer.

7. Any feature of a living thing is called a(n)

   a. instinct.

   b. trait.

   c. inherited trait.

   d. learned trait.

8. Inherited traits come from

   a. parents.

   b. skills.

   c. environment.

   d. offspring.

9. The inside of a seed contains a(n)

   a. egg.

   b. fruit.

   c. larva.

   d. embryo.

10. In which stage does an insect change into an adult?

    a. life cycle

    b. metamorphosis

    c. larva

    d. pupa

11. What must happen before a seed forms in a flower or a cone?

    a. pollination

    b. metamorphosis

    c. heredity

    d. germination

12. Which structure in conifers makes seeds?

    a. flower

    b. fruit

    c. cone

    d. embryo

13. Which stage happens before the pupa stage?

    a. adult

    b. egg

    c. larva

    d. seed

Once Upon a Woodpecker

Read the Unit Literature feature in your textbook.

Write About It

Response to Literature This article tells about special features of woodpeckers that help them survive. What are some special features you have that help you survive? Write about them.

Paragraphs should have a clear topic sentence that directly addresses special features that help students survive. The sentences that follow should provide details for the topic sentence, such as which special features help humans survive. Examples would be hands that help people prepare food needed to survive, or ears that help detect danger. Students should use a closing sentence that wraps up the main idea of the paragraph or restates the topic sentence. Good paragraphs will include correct grammar and mechanics and demonstrate proper transition from one idea to the next.

Living Things in Ecosystems

Complete the concept map with the information you learned about the climates and adaptations in Earth's ecosystems. Some answers have been completed for you.

Ecosystem	Climate and Other Characteristics	Adaptations
desert	dry, warm during the day, cool at night	plants: cactus leaves and stems hold water animals: nocturnal to survive warm days
tropical rain forest	hot and damp all year round	plants: leaves have grooves and "drip tips" animals: climb or fly into trees
temperate forest	cold, dry winters and warm, wet summers	plants: trees lose leaves in winter animals: hibernate during the winter
wetland	land under water most of the year	plants: roots spread out above the water animals: catfish breathe oxygen
ocean	warmer in shallow areas, colder in deeper areas	plants (algae): air bladders animals: gills and fins

© Macmillan/McGraw-Hill

Food Chains and Food Webs

Use your textbook to help you fill in the blanks below.

What is an ecosystem?

1. Plants, animals, water, soil, and sunlight in an area all interact to form a(n) _____ecosystem_____ .

2. In an ecosystem, organisms have different homes, or _____habitats_____ where they live.

3. Plants and animals get food, water, and _____shelter_____ from their habitats.

What is a food chain?

4. To live and grow, every living thing needs _____energy_____ .

5. The system by which organisms get and give energy is a(n) _____food chain_____ .

6. Green plants and algae are _____producers_____ that use the Sun's energy to make food.

7. Organisms that get their energy from eating other organisms are called _____consumers_____ .

8. Decomposers, such as worms and _____bacteria_____ , break down dead plants and animals.

9. Decomposers put _____nutrients_____ into the soil that are used by new plants.

What is a food web?

10. A grasshopper is a(n) _____*herbivore*_____ because it gets its energy from eating plants.

11. A heron is a(n) _____*carnivore*_____ because it gets its energy from eating animals.

12. A bear is an omnivore, which means that it eats plants and _____*animals*_____ .

Why are decomposers important?

13. If ecosystems did not have _____*decomposers*_____ , piles of dead plants and animals would build up.

14. The nutrients that decomposers put back into water or soil help other organisms _____*grow*_____ .

Critical Thinking

15. How do living and nonliving things interact in an ecosystem? Use the terms *sunlight, plants, animals, decomposers,* and *soil* in your answer.

Plants use sunlight to make their own food. Animals eat the

plants for energy. Other animals might eat these animals. When

plants and animals die, decomposers break them down into

nutrients. The nutrients go into the soil and help new plants

grow.

_____

Food Chains and Food Webs

Match the correct letter to its description.

a. predator	**c.** decomposer	**e.** food chain	**g.** habitat
b. consumer	**d.** ecosystem	**f.** food web	**h.** producer

1. ___d___ the living and nonliving things that interact in an environment

2. ___h___ an organism that makes its own food

3. ___e___ shows how energy passes from one organism to another in an ecosystem

4. ___b___ an organism that eats other organisms

5. ___a___ hunts other organisms for food

6. ___c___ an organism that breaks down dead plant and animal material

7. ___f___ shows how food chains are linked together

8. ___g___ the place where an organism lives

Food Chains and Food Webs

Use the words in the box to fill in the blanks.

animals	ecosystem	omnivores
consumers	energy	producers
decomposers	herbivores	Sun

Living things depend on other living things and nonliving things around them. All of these things form a(n) ____ecosystem____ . In an ecosystem, ____energy____ passes from one organism to another. Food chains begin with ____producers____ that make their own food. They use energy from the ____Sun____ .

Organisms that cannot make their own food are called ____consumers____ . Consumers that eat plants are called ____herbivores____ . Carnivores eat other ____animals____ . Plants and animals are eaten by ____omnivores____ . Nutrients from dead organisms are recycled by ____decomposers____ .

Types of Ecosystems

Use your textbook to help you fill in the blanks below.

How do ecosystems differ?

1. An ecosystem's long-term weather conditions are

   called its _____climate_____ .

2. Plants grow well in ecosystems that have _____soil_____ rich in humus.

3. Grasslands, forests, oceans, and ponds have _____different_____ types of plants and animals.

What is a desert?

4. A desert is an ecosystem that gets very little _____rain_____ .

5. Deserts have few plants because the soil is mostly

   _____sand_____ , which does not provide them with enough water to survive.

6. Animals in the desert search for food at _____night_____ when temperatures are cooler.

What is a forest?

7. A(n) _____tropical rain_____ forest is an ecosystem that is warm and rainy all year long.

8. The temperatures and rainfall change each season in

   a(n) _____temperate_____ forest.

What is an ocean?

9. The largest environment on Earth is a body of water

called a(n) _____ocean_____ .

10. In the ocean, a ridge made of tiny animals called

_____coral_____ attracts many fish.

What is a wetland?

11. An ecosystem in which the land is often _____wet_____ and sometimes dry is a wetland.

12. Plants grow well in wetlands because the soil is

full of _____nutrients_____ .

13. Wetlands help prevent _____flooding_____ because they absorb water.

Critical Thinking

14. What are three characteristics that make Earth's ecosystems different from one another? Give an example of each characteristic.

Earth's ecosystems have different climates, types of soil, and

kinds of animals. The desert gets very little rain. The temperate

forest has soil rich in humus. The ocean has fish, sea sponges,

and other animals that do not live anywhere else.

_____

© Macmillan/McGraw-Hill

Types of Ecosystems

Match the words in the box to their definitions below.
Write the correct letter in the space provided.

a. climate	**d.** ocean	**g.** tropical forest
b. desert	**e.** soil	**h.** wetland
c. forest	**f.** temperate	

1. ___h___ an ecosystem in which water covers the soil for most of the year

2. ___a___ the long-term weather conditions of an area

3. ___c___ an ecosystem that has many trees

4. ___f___ the type of forest ecosystem found in North America, Europe, and Asia

5. ___e___ a mixture of broken-down rocks and humus

6. ___b___ an ecosystem that has a dry climate

7. ___d___ the largest environment on Earth

8. ___g___ the type of forest that is hot and damp

Types of Ecosystems

Use the words in the box to fill in the blanks below.

climates	fresh	ocean	soil
desert	hot	seasons	Sun
dry	humus	shallow	tropical rain

Earth's ecosystems differ in many ways. They have

different types of _____climates_____ , soil, plants, and

animals. The largest ecosystem is the _____ocean_____ .

Most ocean organisms live in _____shallow_____ water.

Few animals live deep in the ocean because they would

not get any light from the _____Sun_____ . Wetlands

can have _____fresh_____ water. In a _____desert_____ ,

the land is _____dry_____ and has few plants.

The ecosystem that gets the most rain is the

_____tropical rain_____ forest. It is _____hot_____ there

all year long. The _____soil_____ has few nutrients.

A temperate forest has four _____seasons_____ and soil

that is rich in _____humus_____ . Many animals can be

found in temperate forests.

Meet Ana Luz Porzecanski

Read the Reading in Science feature in your textbook.
Work with a partner to answer the following
questions.

Characteristics of the Tinamou

1. What kind of animal is the tinamou?
How do you know?

It is a bird because it has feathers and lays eggs.

2. What colors is the tinamou?

brown and grey

3. In which type of ecosystem does the tinamou live?

grasslands known as pampas

Characteristics of the Other Animal

Think of an animal that has some things in common
with the tinamou. Answer the questions below.

1. What is the other animal? Describe it.

Possible answer: It is a turtle. It has a shell and scales, and

lays eggs.

2. What color is the animal?

Possible answers: green or brown

3. In which type of ecosystem does it live?

Possible answer: a wetland

Write About It

Compare and Contrast Work with a partner to compare the tinamou with another animal you know about. List ways the animals are alike and different in a Venn diagram. Then use your diagram to write about the animals.

Use your answers to the questions on the previous page to fill in the Venn diagram.

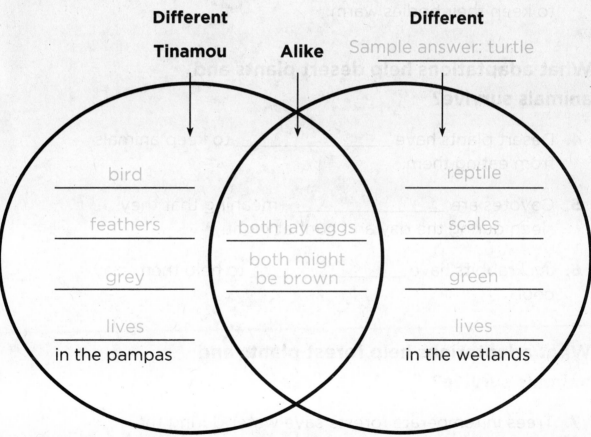

Different   **Different**

Tinamou   **Alike**   Sample answer: turtle

bird   reptile

feathers   both lay eggs   scales

grey   both might   green
            be brown

lives   lives

in the pampas   in the wetlands

1. On a separate piece of paper, explain how the two animals are alike and different.

   Possible answers: Both lay eggs. A tinamou has feathers.

© Macmillan/McGraw-Hill

Adaptations

Use your textbook to help you fill in the blanks below.

How are living things built to survive?

1. A frog's sticky tongue is a(n) ___adaptation___ that helps it find food.

2. Some animals escape from danger by using ___camouflage___ to blend in with their environment.

3. In cold climates, some animals have ___blubber___ to keep their bodies warm.

What adaptations help desert plants and animals survive?

4. Desert plants have ___spines___ to keep animals from eating them.

5. Coyotes are ___nocturnal___ , meaning that they sleep during the day and hunt at night.

6. Jackrabbits have ___large ears___ to help them stay cool.

What adaptations help forest plants and animals survive?

7. Trees in temperate forests save water during the

   winter by losing their ___leaves___ .

8. An organism that imitates another is using ___mimicry___ .

© Macmillan/McGraw-Hill

9. Some animals _____ during the winter.
    hibernate

What adaptations help ocean plants and animals survive?

10. Plant-like organisms that live in the ocean and make

    their own food are called _____ .
    algae

11. Algae without roots have _____ to help
    air bladders
    them float on the water's surface.

What are adaptations to a wetland?

12. Wetland plants have adaptations to help them

    survive when _____ change.
    water levels

13. During dry seasons, wetland catfish breathe _____
    oxygen
    from the air.

Critical Thinking

14. What are three ways in which adaptations help
    organisms survive? Give an example of each.

    Adaptations help organisms find food, escape danger, and live

    in different climates. Frogs have sticky tongues to catch insects.

    Some snakes use camouflage so that they will not be seen.

    Some animals in cold climates keep warm through a layer

    of blubber.

Adaptations

Use the words in the box to fill in the blanks below.

adaptation	blubber	hibernate	mimicry
air bladders	camouflage	migrate	nocturnal

1. A special characteristic that helps an organism survive in its environment is a(n) _____adaptation_____ .

2. Animals that are active at night are _____nocturnal_____ .

3. A layer of fat under the skin of some animals who live in cold climates is _____blubber_____ .

4. When animals move from one place to another, they _____migrate_____ .

5. Blending into one's environment is called _____camouflage_____ .

6. When one living thing imitates another in color or shape, it is using _____mimicry_____ .

7. Animals go into a deep sleep when they _____hibernate_____ .

8. Balloon-like structures that help algae float are called _____air bladders_____ .

Adaptations

Use the words in the box to fill in the blanks below.

air bladders	gills	nocturnal
angler fish	hibernate	roots
camouflage	migrate	sunlight

Living things have adaptations that help them find

food and water. Desert plants have long _____roots_____

to find water. Algae use _____air bladders_____ to float on

the ocean's surface to get _____sunlight_____ . Deep

in the ocean, the _____angler fish_____ has a light that

attracts its food.

Some animals use _____camouflage_____ to blend into

their environment. Animals in the desert are _____nocturnal_____ ,

or active at night. Some forest animals _____hibernate_____

during the winter when food is hard to find. Fish have

_____gills_____ so that they can breathe underwater.

Some animals _____migrate_____ when the seasons change.

Plant and animal adaptations can be found in every

environment.

Living Things in Ecosystems

Circle the letter of the best answer.

1. All the living and non-living things in an environment form a(n)

   a. habitat.

   b. food web.

   c. ecosystem.

   d. food chain.

2. Which ecosystem has the highest density of living things?

   a. temperate forest

   b. tropical rain forest

   c. wetland

   d. desert

3. To avoid very cold temperatures when the seasons change, some animals

   a. use mimicry.

   b. use camouflage.

   c. become nocturnal.

   d. migrate.

4. Which type of organism makes its own food?

   a. consumer

   b. producer

   c. herbivore

   d. omnivore

5. Which of these ecosystems receives the least rain?

   a. forest

   b. ocean

   c. desert

   d. wetland

6. An animal that is active at night is said to be

   a. hibernating.

   b. migrating.

   c. using camouflage.

   d. nocturnal.

Circle the letter of the best answer.

7. Which of the following is an example of a consumer?

   a. plant

   b. insect

   c. algae

   d. bacteria

8. Which of the following shows that organisms usually eat more than one type of food?

   a. habitat

   b. ecosystem

   c. food chain

   d. food web

9. Which ecosystem has soil that is under water most of the year?

   a. desert

   b. wetland

   c. temperate forest

   d. tropical rain forest

10. Which type of organism breaks down dead plants and animals?

   a. decomposer

   b. carnivore

   c. producer

   d. omnivore

11. What do some animals do so that they can survive with less energy in winter?

   a. use camouflage

   b. use mimicry

   c. migrate

   d. hibernate

12. Climate describes an ecosystem's

   a. weather.

   b. plants.

   c. soil.

   d. nonliving things.

© Macmillan/McGraw-Hill

Changes in Ecosystems

Complete the concept map with the information that you learned about changes in ecosystems.

Living things change ecosystems.

► The changes can be small or ___large___ .

► Living things can cause changes when they compete for ___resources___ .

► When people ___pollute___ land, air, or water, they cause changes in ecosystems.

► People also cause changes by clearing ___land___ for towns and cities.

Changes in Ecosystems

Changes in ecosystems affect living things.

► Changes that affect living things include ___floods___ , ___droughts___ , and disease.

► Living things that cannot adjust to changes may become ___endangered___ , which means that only a few of their population remain.

Fossils tell about past changes.

► People study ___fossils___ to learn about ancient organisms and changes on Earth over time.

► Dinosaurs are ___extinct___ , possibly because of a meteor, and ___saber-toothed cats___ are extinct, possibly because of climate changes.

© Macmillan/McGraw-Hill

Living Things Change Their Environments

Use your textbook to help you fill in the blanks.

How do living things change their environments?

1. Living things change their ___environments___ in small and large ways.

2. One way that plants change the environment is by absorbing ___water___ from the soil.

3. Worms change the environment by adding ___nutrients___ to the soil.

4. When water is limited, a struggle, or ___competition___, among plants may occur.

5. Food and water are ___resources___ that living things need to survive.

How do people change their environments?

6. The organisms that cause the most changes to the environment are ___people___ .

7. Plants and animals can lose their ___homes___ when people clear forests.

8. Cars, factories, and trash can harm the environment by causing ___pollution___ .

9. Pollution increases when ___wetlands___ are drained.

© Macmillan/McGraw-Hill

10. An organism that is new to an environment can harm

it by competing for _limited resources_ that other
plants and animals need.

How can people protect their environments?

11. You protect the environment when you ___reduce___
the amount of paper you use.

12. You protect the environment when you ___reuse___
newspapers to line pet cages.

13. Businesses protect the environment when they

___recycle___ old newspapers into new paper
products.

14. When you plant a tree, you help keep ___soil___
from washing away.

Critical Thinking

15. How do you help plants and animals when you
reduce, reuse, and recycle paper products?

Possible answer: Paper is made from trees. When you practice

the 3 Rs, less paper has to be made. Fewer trees have to be cut

down for new paper products. This saves forests from being

cleared, and plants and animals will not lose their homes.

_____

_____

© Macmillan/McGraw-Hill

Living Things Change Their Environments

**Match the word with its correct description below.
Write its letter in the space provided.**

a. competition	**c.** predators	**e.** reduce	**g.** reuse
b. pollution	**d.** recycle	**f.** resource	

1. ____d____ to turn old things into new things

2. ____b____ when harmful materials get into the air,
land, or water

3. ____c____ animals that hunt other animals

4. ____f____ something that helps an organism survive

5. ____g____ to use something again

6. ____a____ the struggle for survival among living things

7. ____e____ to use less of something

Living Things Change Their Environments

Use the words in the box to fill in the blanks below.

competition	healthy	reuse
environment	homes	soil
forests	recycle	wetlands

To meet their needs, living things, including people, change the environment. To build towns and cities, people sometimes drain _____wetlands_____ and clear _____forests_____ . This takes _____homes_____ away from many plants and animals. Another cause of change is _____competition_____ for resources, such as water and space, among living things. As a result, the environment is changed.

People can improve the _____environment_____ . We can produce less trash if we _____reuse_____ and _____recycle_____ . We can keep the environment _____healthy_____ by planting trees. Trees help keep _____soil_____ in its place. They can also help clean the air.

© Macmillan/McGraw-Hill

Changes Affect Living Things

Use your textbook to help you fill in the blanks.

What are some ways environments change?

1. An environment can change when a(n) _____ flood _____ covers dry land with water.

2. Living things are harmed when they do not get

   enough water during a(n) _____ drought _____ .

3. Floods and droughts are types of _____ natural disasters _____ .

4. Animals can lose their homes when lightning starts

   a(n) _____ wildfire _____ .

5. Some bacteria and mold can cause _____ diseases _____
   that harm many living things.

How do organisms respond to changes?

6. Burrowing in the mud is a(n) _____ adaptation _____
   that helps frogs survive in a dry environment.

7. Some animals that cannot survive in a changed

   environment may move to a new _____ habitat _____ .

8. If organisms cannot move and the environment

   has changed too much, the organisms will _____ die _____ .

How do environmental changes affect an entire community?

9. In the grasslands of the central United States,

prairie dogs build _____tunnels_____ and eat _____grasses_____ .

10. Prairie dogs are food for _____eagles_____ and

_____coyotes_____ .

How does a living thing become endangered?

11. The population of a(n) _____endangered_____ organism is very small.

12. Because it cannot adjust to dry conditions, the

_____Saharan cypress_____ may disappear.

13. People can cause organisms to become endangered

when they _____hunt_____ them.

Critical Thinking

14. Would animals in a forest be harmed if a disease spread that only affected plants? Explain why or why not.

_____Possible answer: Yes, animals would be harmed. Some animals_____

_____eat plants for food, and other animals eat them. Therefore, many_____

_____animals may not have enough food. The animals would also lose_____

_____their places to live and hide._____

© Macmillan/McGraw-Hill

Changes Affect Living Things

Match the word with its correct description below.
Write the letter of the word in the space provided.

---

a. community	**d.** endangered	**g.** natural disaster
b. disease	**e.** flood	**h.** wildfire
c. drought	**f.** population	

---

1. ____c____ a long period of time with no rain

2. ____e____ something that covers dry land with water

3 ____b____ can be caused by bacteria or mold

4. ____h____ can start when lightning strikes a dry area

5. ____a____ all of the organisms in an ecosystem

6. ____g____ a flood is an example of this

7. ____d____ when an organism has only a few living members of its population left

8. ____f____ all of the members of a single type of organism in an ecosystem

Name _____ Date _____

Changes Affect Living Things

Use the words in the box to fill in the blanks below.

disease	migrate	plants	tunnels
floods	organism	prairie dogs	water

Changes in an environment affect its living things.

A change that affects only one type of ___organism___

can eventually affect other populations. For example,

coyotes eat ___prairie dogs___ . Mice and snakes live in

the ___tunnels___ that prairie dogs build. If a(n)

___disease___ destroyed the prairie dogs, all of

these other animals would be affected.

Besides diseases, natural disasters such as

___floods___ and droughts can change an

environment. When dry land is covered by water, soil

and ___plants___ can be washed away. Some

organisms die from too little ___water___ during

a drought. Organisms must ___migrate___ or adjust

to a changing environment. If they do not, they may

become endangered.

© Macmillan/McGraw-Hill

Save the Koala Bears

Read the Writing in Science feature in your textbook.

Write About It

Persausive Writing Choose an endangered animal you care about. Research to find out why this animal is in trouble. Write a paragraph to convince readers that this animal should be saved. Be sure to end with a strong argument.

Getting Ideas

Fill out the chart below. Write your opinion about your endangered animal in the top oval. Write down the reasons that support your opinion in the bottom ovals.

Opinion

Possible answer: We need to protect the giant panda bear before it dies out.

Reason

Only 1,600 panda bears are left in the wild.

Reason

If these creatures die out, it would affect the balance of nature.

© Macmillan/McGraw-Hill

Name _____ Date _____

Planning and Organizing

Isabella wrote about the giant panda bear. Does her sentence tell why we should protect the panda? If so, write "yes." Write "no" if it does not.

1. If pandas die out, it will affect the balance of nature. ___yes___

2. I saw a beautiful panda bear in the zoo. ___no___

Drafting

Pick an animal. Write a sentence that states your opinion about saving it.

Possible sentence: We need to protect the giant panda bear.
_____

Now write your paragraph on a separate piece of paper. Begin with the sentence that you wrote above.

Revising and Proofreading

Here are some sentences that Isabella wrote. Proofread them. Find the five spelling errors. Cross out each misspelled word. Write the correct spelling above it.

Panda bears have lived in bamboo forests for milions of years. If the jiant panda bear dyes out, the Earth will lose one of the most beautiful kreatures in the world. I beleive that people must take action now.

Now revise and proofread your writing.
Ask yourself:

▶ Did I state my opinion about an endangered animal?

▶ Did I include convincing reasons?

▶ Did I correct all mistakes?

© Macmillan/McGraw-Hill

Living Things of the Past

Use your textbook to help you fill in the blanks.

What can happen if the environment suddenly changes?

1. People know about organisms that lived long ago

   because of remains called _____fossils_____ .

2. A type of organism that has no living population is

   said to be _____extinct_____ .

3. Large animals called _____saber-toothed cats_____ became
   extinct about 10,000 years ago when the climate
   changed.

4. Ice covered much of Earth during the _____Ice Age_____ .

5. Disease and dry weather caused the _____St. Helena Olive tree_____
   to become extinct in 2004.

How can we learn about things that lived long ago?

6. Scientists can tell what animals ate by studying

   their _____teeth_____ .

7. Scientists learn how animals moved by studying

   their _____bones_____ .

8. Fish fossils found on _____land_____ teach us that

the area was once covered by _____water_____ .

9. The _____oldest_____ fossils are usually deep below
the surface of the ground.

10. The _____youngest_____ fossils are usually below
ground but close to the surface.

How are living things of today similar to those that lived long ago?

11. Fossils do not show how organisms used their _____body parts_____ .

12. Elephants today are similar to _____woolly mammoths_____ that
lived long ago.

13. The pterodactyl was a flying reptile that used its beak

and claws to catch fish, just as the _____eagle_____
does today.

Critical Thinking

14. Tropical plants can be found where it is hot and
rainy. Fossils of tropical plants have been found in
a place where it is cold today. What can you infer
from this finding?

Possible answer: The place where it is cold today probably used

to have a hot and rainy climate.

_____

_____

Living Things of the Past

Who am I? What am I?

Choose a word from the box below that answers each question and write its letter in the space provided.

a. extinct	**d.** saber-toothed cat
b. fossil	**e.** St. Helena Olive tree
c. Ice Age	**f.** woolly mammoth

1. _____b_____ I am the remains of an organism that lived long ago. What am I?

2. _____f_____ I am an ancient animal that used a trunk as elephants do today. Who am I?

3. _____a_____ I was a living organism, but there are no more of my kind alive. What am I?

4. _____d_____ I am a big animal that became extinct when the climate changed thousands of years ago. Who am I?

5. _____e_____ I am a type of tree that is extinct because of disease and dry weather. What am I?

6. _____c_____ During my time, large ice sheets covered much of the land. What am I?

Living Things of the Past

Use the words in the box to fill in the blanks below.

ate	extinct	similar
body parts	layers	woolly mammoths
Earth	meteor	

Scientists learn about ancient organisms by studying fossils. They learn how animals looked, how they moved, and what they ____ate____ . Some living things today look ____similar____ to organisms of long ago. Scientists can infer from them how ancient organisms used their ____body parts____ . For example, elephants look like ____woolly mammoths____ .

From fossils, scientists also learn how ____Earth____ has changed over time. They find fossils in its rock ____layers____ . Scientists think that some animals are ____extinct____ because of natural events. For example, dinosaurs may have died when a(n) ____meteor____ hit Earth. Other animals became extinct because of humans' activities, competition, and disease.

© Macmillan/McGraw-Hill

Looking at Dinosaurs

Read the Reading in Science feature in your textbook.

Write About It

Fact and Opinion What animal do you think dinosaurs are like? What animal do scientists think dinosaurs are like? Why do scientists think this?

Planning and Organizing

Answer the following questions.

What animal do you think dinosaurs are like?

Possible answer: Dinosaurs are like tortoises because they are both

reptiles and have scaly skin.

_____

_____

What animal do scientists think dinosaurs are like?
Why do scientists think this?

Possible answer: Most scientists agree that dinosaurs are like birds

because they laid eggs and some were covered in feathers. They

think this based upon evidence that was discovered.

_____

Drafting

Write a paragraph explaining how one of the previous answers is an opinion and the other is a fact. Use examples of dinosaur discoveries to support your writing.

Possible answer: The first answer is my opinion because it is what I thought about dinosaurs without any evidence. The second answer is fact because it is supported by scientific evidence, such as the discovery of dinosaur nests containing eggs in the Gobi Desert in China and dinosaur fossils with feathers.

_____

_____

_____

_____

_____

_____

_____

_____

_____

_____

_____

Changes in Ecosystems

Circle the letter of the best answer.

1. To use something again is to

   a. reuse.

   b. reduce.

   c. recycle.

   d. compete.

2. Which is an example of a resource?

   a. fossil

   b. air

   c. disease

   d. adaptation

3. Soil and plants may be washed away during a

   a. drought.

   b. wildlife.

   c. flood.

   d. disease.

4. When you use less of something, you

   a. adapt.

   b. recycle.

   c. reuse.

   d. reduce.

5. Which organism is extinct?

   a. lizard

   b. eagle

   c. elephant

   d. woolly mammoth

6. In the grasslands of the central United States, which organism is prey for a coyote?

   a. grass

   b. prairie dog

   c. hawk

   d. eagle

Circle the letter of the best answer.

7. For which of these resources do plants compete?

   a. prey

   b. food

   c. sunlight

   d. air

8. Scientists learn the most about ancient organisms by studying

   a. fossils.

   b. resources.

   c. natural disasters.

   d. adaptations.

9. Which does a business do when it makes new cans from old cans?

   a. adapt

   b. reuse

   c. reduce

   d. recycle

10. Which is a natural disaster caused by too little rain?

   a. pollution

   b. disease

   c. drought

   d. flood

11. What is produced when harmful things are put in the air, water or on land?

   a. pollution

   b. competition

   c. extinction

   d. adaptation

12. Which is a predator?

   a. tree

   b. fungus

   c. wild horse

   d. hawk

One Cool Adventure

Read the Unit Literature feature in your textbook.

Write About It

Response to Literature This article tells about the first women to cross Antarctica on skis. What do you know about Antarctica or other places on Earth? Write about an imaginary trip around the world. What kinds of things might you see? Write about it.

Each paragraph should have a clear topic sentence that directly

addresses the kinds of things students might experience on a trip

around the world. The sentences that follow the topic sentence

should include details about the trip, such as the humidity they

would encounter while trekking in the jungles of Brazil. Students

should use a closing sentence that restates the main idea of the

paragraph. Good paragraphs will contain vivid words and imagery,

include correct grammar and mechanics, and demonstrate proper

transition from one idea to the next.

_____

_____

_____

_____

Name _____ Date _____

Earth Changes

Complete the concept map about Earth's features and how they can change. Some parts have been done for you.

Some of Earth's Features

1. mountains

2. _____ oceans

3. _____ valleys

4. plateaus

5. _____ islands

any others students
might choose

6. _____

Some things cause landforms on Earth's crust to change suddenly.

1. volcanoes

2. _____ earthquakes

3. floods

4. _____ landslides

Some things cause landforms on Earth's crust to change very slowly.

1. Weathering by:

a. moving water

b. _____ wind

c. ice

d. _____ plant roots

2. Erosion by:

a. _____ wind

b. _____ water

c. glaciers

© Macmillan/McGraw-Hill

Earth's Features

Use your textbook to help you fill in the blanks.

What covers Earth's surface?

1. More than half of Earth is covered

   by ___water___ .

2. Most of Earth is covered by ___oceans___ ,
   which are made up of salt water.

3. Rivers and glaciers are made up of ___fresh___
   water.

4. Water that is not ___salty___ is fresh water.

5. Earth's ___continents___ make up seven great
   land areas.

What are some of Earth's land and water features?

6. A deep, narrow valley with steep sides is

   a(n) ___canyon___ .

7. A landform with water all around it is

   a(n) ___island___ .

8. Rivers are bodies of ___moving___ water.

9. Land that is flat on top and higher than the land

   around it is called a(n) ___plateau___ .

What land features are in the oceans?

10. The land under an ocean at the edge of a continent is called a <u>continental shelf</u>.

11. Land that stretches for thousands of miles across the ocean is called the <u>abyssal plain</u>.

12. Canyons called <u>trenches</u> form the deepest parts of the ocean floor.

What are the layers of Earth?

13. Earth's <u>crust</u> makes up the continents and the ocean floor.

14. Earth's crust is a(n) <u>thin</u>, cool layer.

15. Under the crust is a layer called the <u>mantle</u>.

16. The deepest and hottest layer of Earth is the <u>core</u>.

17. The outer core is made up of <u>melted</u> rock.

18. Earth's inner core is made up of <u>solid</u> rock.

Critical Thinking

19. What can a map show you about Earth's features?

<u>A map can show that Earth has land and water features. It can show that most of Earth is covered by water.</u>

Earth's Features

Match each word with its definition.

a. abyssal plain	**e.** core	**i.** ocean
b. coast	**f.** crust	**j.** trench
c. continent	**g.** island	
d. continental shelf	**h.** mantle	

1. _____i_____ a large body of salt water

2. _____f_____ Earth's thin outer layer

3. _____c_____ a great area of land

4. _____h_____ the layer immediately below Earth's crust

5. _____b_____ land that borders the ocean

6. _____j_____ a canyon that is the deepest part of the ocean floor

7. _____e_____ the deepest and hottest layer of Earth

8. _____g_____ land with water all around it

9. _____d_____ a plateau under the ocean at the edge of a continent

10. _____a_____ a deep, flat part of the ocean floor, thousands of kilometers wide

Earth's Features

Fill in the blanks.

abyssal plain	landforms	plain
continental shelf	mantle	plateau
crust	ocean	

The seven large land areas of Earth are the

continents. Continents have ___landforms___ such as

mountains and valleys. A high, flat landform with steep

sides is a(n) ___plateau___ . Another landform is

a(n) ___plain___ , which is flat and wide.

The outer layer of Earth is the ___crust___ . It

is thin and cool. The layer just below the crust is the

___mantle___ . It is made up of rock that is hot

and flowing.

Most of Earth is covered by salty ___ocean___

water. Land under the ocean along a coast forms the

___continental shelf___ . Farther out, the wide, flat ___abyssal plain___

makes up the ocean floor. A deep canyon in the ocean floor is

called a trench.

© Macmillan/McGraw-Hill

Sudden Changes to Earth

Use your textbook to help you fill in the blanks.

What are earthquakes?

1. The huge rocks that make up Earth's crust

   can _____move_____ .

2. Rocks below ground can bend when they

   _____slide past_____ each other.

3. Rocks below ground can press against other rocks

   and _____pull apart_____ .

4. Rocks that bend can _____snap back_____ and cause
   sudden movement.

5. A sudden movement of rocks in Earth's crust is

   an _____earthquake_____ .

6. An earthquake can be _____weak_____ or
   very strong.

7. The land _____vibrates_____ , or shakes, during
   an earthquake.

8. During an earthquake, vibrations travel as

   _____waves_____ in all directions.

What are volcanoes?

9. A mountain around an opening in Earth's crust is

   a _____volcano_____ .

10. Melted rock is called _____magma_____ when it is in Earth's crust and mantle.

11. Melted rock that flows through an opening in the crust is called _____lava_____ .

12. A volcanic _____mountain_____ forms when lava, rocks, and ash pile up in layers.

13. Lava can _____ooze_____ from a volcano, or it can explode out of it.

What are landslides and floods?

14. The force that pulls on all objects, including rocks, is called _____gravity_____ .

15. When rocks and soil move downhill very fast, a _____landslide_____ occurs.

16. A river that overflows can cause a _____flood_____ on land that is usually dry.

Critical Thinking

17. How can earthquakes, volcanoes, landslides, and floods change the shape of coastlines?

Possible answer: Earthquakes, floods, and landslides can make

land fall into the ocean and change the shape of the coast.

Volcanic mountains under the ocean can rise above the surface

of the sea and create islands.

© Macmillan/McGraw-Hill

Sudden Changes to Earth

Match the correct word to its definition by writing its letter in the space provided.

a. earthquake	**c.** gravity	**e.** lava	**g.** vibrations
b. flood	**d.** landslide	**f.** magma	**h.** volcano

1. ___g___ shaking felt during an earthquake

2. ___f___ melted rock in the crust and mantle

3. ___c___ the force that pulls objects downward

4. ___a___ sudden movement of rocks in Earth's crust

5. ___b___ water that flows over land that is normally dry

6. ___e___ melted rock that flows through an opening and out onto land

7. ___d___ rock and soil pulled down a hill by gravity

8. ___h___ a mountain that builds up in Earth's crust around an opening

Name _____ Date _____

Sudden Changes to Earth

Use the words in the box to fill in the blanks.

crust	flood	magma
earthquake	lava	volcano

The land that makes up Earth's surface can change very quickly. Large, flat rocks in Earth's _____crust_____ can slide or press against each other. If they break or snap back, they cause an _____earthquake_____ . Water can also change land. Sometimes heavy rain fills a river, and water flows over the banks. A _____flood_____ forms as water flows onto land that is usually dry.

Hot, melted rock in Earth's crust is called _____magma_____ . Magma that flows out through an opening in the crust is called _____lava_____ . A mountain that forms around an opening in Earth's crust is a _____volcano_____ . Lava can ooze or explode from a volcano and change the land.

Slide on the Shore

Read the Reading in Science feature in your textbook.

Cause and Effect

Use the graphic organizer to list the causes and
effects of erosion and ways to prevent it.

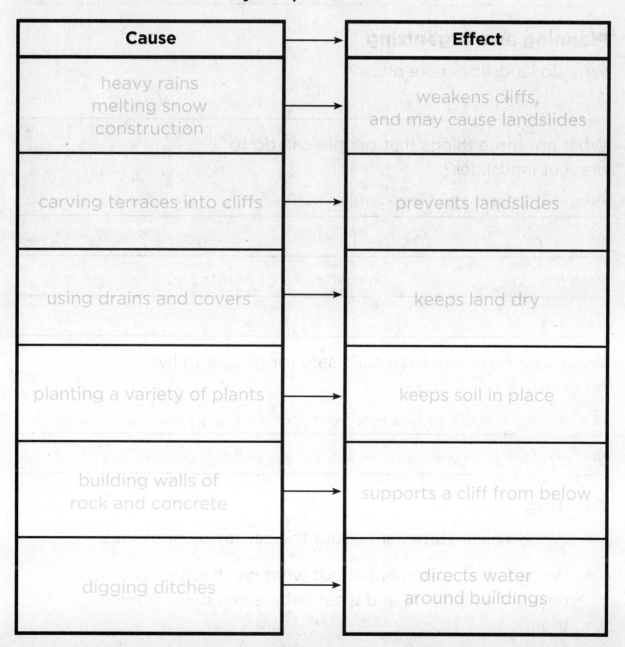

Cause	Effect
heavy rains melting snow construction	weakens cliffs, and may cause landslides
carving terraces into cliffs	prevents landslides
using drains and covers	keeps land dry
planting a variety of plants	keeps soil in place
building walls of rock and concrete	supports a cliff from below
digging ditches	directs water around buildings

© Macmillan/McGraw-Hill

Write About It

Cause and Effect Read the article again with a partner. Write a few sentences that tell what causes landslides to happen. Include also what people can do to prevent them from happening.

Planning and Organizing

Why do landslides take place?

Heavy rains, melting snow, and construction weaken cliffs.

What are three things that people can do to prevent landslides?

People can carve terraces into cliffs so that rocks and water do not

flow to the bottoms of cliffs. They can use drains and covers to keep

land dry. They can also plant a variety of plants to keep soil in place.

What structures can keep cliffs safe for people to live on or near?

Walls of concrete or rock can be built to support a cliff from below

and ditches can be dug to direct water around buildings.

Drafting

▶ Write a clear statement about the causes of landslides.

▶ Write a clear statement about what can happen as a result of a landslide and what people can do to prevent them.

© Macmillan/McGraw-Hill

Weathering and Erosion

Use your textbook to help you fill in the blanks.

What is weathering?

1. The process of ____weathering____ breaks down rocks into smaller and smaller pieces.

2. Weathering can break down rocks into ____sand____ and soil.

3. Weathering can be caused by ____running water____, wind, rain, and ice.

4. Rocks can weather when they ____scrape____ against each other.

5. Water that freezes in a crack ____expands____ and makes the crack larger.

6. Over time, freezing and ____thawing____ water breaks down a rock into smaller pieces.

7. Plant ____roots____ grow into cracks in a rock and split the rock.

What is erosion?

8. Pieces of weathered rock get moved to other places by ____erosion____.

9. Erosion can happen when ____gravity____ pulls weathered rock and soil downhill.

10. One cause of erosion is the ___moving water___ in rivers and the ocean.

11. Rock can be dropped or ___deposited___ by wind when the wind slows down.

12. Rocks are carried along inside a(n) ___glacier___ and drop off as it melts.

How can people change the land?

13. Digging a(n) ___hole___ is a small example of how people can change land.

14. Land is changed when trees are cut to build ___roads___, ___stores___, and ___homes___.

15. Soil can wash away if trees are not ___replanted___.

16. Land can change when it is ___dug up___ for valuable minerals and metals.

Critical Thinking

17. Why would planting trees help stop erosion caused by wind and water?

Planting trees would help stop wind erosion because it would

keep the wind from blowing soil away. Trees also help keep

water from washing soil away.

_____

Weathering and Erosion

What am I?

Choose a word from the word box below that answers each question and write the correct letter in the space provided.

a. erosion	**c.** gravity	**e.** soil	**g.** wind
b. glacier	**d.** roots	**f.** weathering	

1. I am a force that pulls materials downhill.

   What am I? ___c___

2. I drop or deposit small bits of rock when I slow down.

   What am I? ___g___

3. I can grow in small cracks and split rocks apart.

   What am I? ___d___

4. Rocks freeze under me. Then I drop them in new

   places when I begin to melt. What am I? ___b___

5. I am the movement of weathered rock.

   What am I? ___a___

6. I cause rocks to break down with the help of running

   water, rain, and ice. What am I? ___f___

7. I am the result of rocks, being broken down into

   smaller pieces. What am I? ___e___

Weathering and Erosion

Fill in the blanks.

blow away	erosion	freezing	scrape
deposited	expands	plant roots	wind

The weathering of rocks is the way that rocks get broken into small pieces. Weathered rock moving from one place to another is ___erosion___ . Sand and rocks are picked up and ___deposited___ in a new place. Moving water, glaciers, and ___wind___ help erosion take place. People can cause erosion by cutting down trees, which can cause the soil to ___blow away___ .

Some weathering happens because water ___expands___ when it freezes in a crack in a rock. Repeated ___freezing___ and thawing helps break the rock. Some rocks weather when they ___scrape___ against other rocks. Rocks also break apart when ___plant roots___ grow into their cracks. They also break when gravity pulls rocks downhill.

© Macmillan/McGraw-Hill

Missing Noses

Read the Writing in Science feature in your textbook.

 Write About It

Expository Writing Write a paragraph to describe other causes of weathering. Remember to start with a topic sentence and to end with a conclusion.

Getting Ideas

Water is one cause of weathering and erosion. What are some other causes? Write them in the chart below.

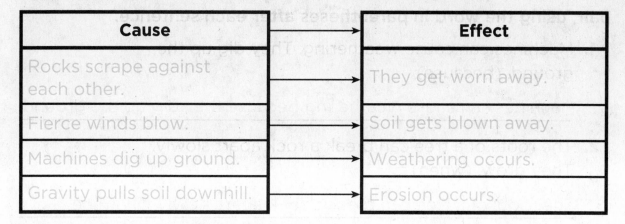

Cause	Effect
Rocks scrape against each other.	They get worn away.
Fierce winds blow.	Soil gets blown away.
Machines dig up ground.	Weathering occurs.
Gravity pulls soil downhill.	Erosion occurs.

Planning and Organizing

Here are three sentences that Suki wrote. Write "MI" if the sentence tells a main idea. Write "D" if it gives a detail.

1. _____D_____ Large rocks get worn away by scraping against smaller rocks.

2. _____D_____ Wind can cause the soil to erode.

3. _____MI_____ There are several causes of weathering and erosion.

Drafting

Write a topic sentence. Make sure it states your main idea about weathering and erosion.

Possible sentence: There are several causes of weathering and

erosion.

Now write your paragraph. Use a separate piece of paper. Begin with your topic sentence. Then write about other causes of weathering and erosion. Reach a conclusion at the end.

Revising and Proofreading

Here are some sentences Suki wrote. Combine each pair, using the word in parentheses after each sentence.

1. Machines can cause weathering. They dig up the ground. (because)

Machines can cause weathering because they dig up the ground.

2. The roots of a tree can break a rock apart slowly. They grow. (when)

The roots of a tree can break a rock apart slowly when they

grow.

3. Rocks will weather. An animal digs them up. (after)

Rocks will weather after an animal digs them up.

Now revise and proofread your writing. Ask yourself:

▶ Did I begin with a topic sentence?

▶ Did I include facts and details?

▶ Did I correct all mistakes?

© Macmillan/McGraw-Hill

Earth Changes

Circle the letter of the best answer.

1. The middle layer of Earth is the
   a. crust.
   b. mantle.
   c. inner core.
   d. outer core.

2. The continents and ocean floor make up Earth's
   a. mantle.
   b. coast.
   c. crust.
   d. core.

3. What forms when water suddenly flows over land that is usually dry?
   a. an ocean
   b. a volcano
   c. a landslide
   d. a flood

4. A continental shelf is flat and slopes off into the ocean. It is similar to a
   a. plateau.
   b. coast.
   c. plain.
   d. trench.

5. Most of Earth is covered by
   a. sand and soil.
   b. oceans.
   c. lakes.
   d. melted rock.

6. When water freezes, it
   a. flows.
   b. expands.
   c. shrinks.
   d. melts.

© Macmillan/McGraw-Hill

Circle the letter of the best answer.

7. What change occurs when rocks are carried to a new place?

   a. erosion

   b. melting

   c. freezing and thawing

   d. formation of a new rock

8. Melted rock that flows out onto land is called

   a. magma.

   b. lava.

   c. a landslide.

   d. a slab of rock.

9. An earthquake occurs when slabs of rock in the Earth's crust

   a. move slowly.

   b. explode.

   c. move suddenly.

   d. ooze onto land.

10. An earthquake can happen because rocks in Earth's crust can

    a. melt and harden.

    b. bend and snap back.

    c. freeze and thaw.

    d. ooze and explode.

11. Gravity causes erosion on a hillside when it

    a. pushes rocks along.

    b. holds rocks in place.

    c. pulls rocks downhill.

    d. bends rocks.

12. What makes a slowing wind drop pieces of weathered rock?

    a. moving water

    b. gravity

    c. glaciers

    d. animals

13. The deepest part of an ocean is its

    a. abyssal plain.

    b. coast.

    c. continental shelf.

    d. trench.

Using Earth's Resources

Complete the concept map with the information you learned about Earth's resources.

Minerals and Rocks

The three types of rock

are _____igneous_____,

_____metamorphic_____, and

_____sedimentary_____. Rocks

and minerals are used to

make _____jewelry,_____

_____pencils, salt,_____

_____roads, and houses_____.

Soil

Soil is made of

_____minerals, humus,_____

_____weathered rock,_____

_____water, air, and_____

_____living things_____.

It provides a place
for plants to

_____grow_____.

Earth's Resources

Air and Water

Air is important because
it has

_____oxygen_____ that

_____animals_____ need.

People need water

for _____drinking and_____

_____cooking_____.

Fossils and Fuels

Some types of fossils are

_____bones_____,

stony models

_____molds_____, and

_____casts_____. Fossil

fuels are a source

of _____energy_____.

Minerals and Rocks

Use your textbook to help you fill in the blanks.

What are minerals?

1. Solid, nonliving substances called ____minerals____ are found in rocks and soil.

2. It is possible to tell one mineral from another because minerals have their own ____properties____ .

3. Minerals cannot be identified by ____color____ alone because some minerals come in many colors.

4. One property of minerals is the color of their powder, or ____streak____ .

5. A mineral's ____luster____ can be described by the way light bounces off of it.

6. Minerals are scratched in order to investigate the property called ____hardness____ .

What are rocks?

7. A rock with large grains has a coarse ____texture____ .

8. A rock that forms from magma or lava is classified as a(n) ____igneous rock____ .

9. Granite that has formed from melted rock inside Earth is called ____magma____ .

© Macmillan/McGraw-Hill

10. Basalt that has formed from melted rock on Earth's

surface is called _____lava_____ .

What are sedimentary and metamorphic rocks?

11. A rock that forms from layers of sediment is classified

as a(n) _____sedimentary rock_____ .

12. Another name for the tiny bits of rock that make up

shale is _____sediment_____ .

13. Heating and squeezing rocks inside Earth can form a

kind of rock called ____metamorphic rock____ .

How do we use minerals and rocks?

14. Minerals called _____gems_____ are valued for
their beauty.

15. People make cement from _____limestone_____ and

burn _____coal_____ for fuel.

Critical Thinking

16. Choose three rocks or minerals mentioned in the
textbook that you would use to make a necklace.
Explain your choices based upon their qualities.

Possible answer: I would use the mineral turquoise because of its

pretty color, the rock granite because of its coarse texture, and

the mineral gold because of its color and common use in jewelry.

Minerals and Rocks

What am I?

Choose a word from the box that answers each question below.

a. igneous rock	**c.** metamorphic rock	**e.** sediment
b. luster	**d.** mineral	**f.** sedimentary rock

1. I am the property of a mineral that describes how light reflects from the mineral. What am I? ___b___

2. I am tiny bits of animals, plants, or weathered rock. What am I? ___e___

3. I am a solid, nonliving substance found in nature. What am I? ___d___

4. I formed when layers of sediment piled up and were pressed together. What am I? ___f___

5. I formed deep inside Earth. I was heated and squeezed by the weight of rocks above me. What am I? ___c___

6. I formed when melted rock cooled and hardened, either inside Earth or on Earth's surface. What am I? ___a___

© Macmillan/McGraw-Hill

Minerals and Rocks

Use the words in the box to fill in the blanks below.

animals	lava	metamorphic	sedimentary
hardness	luster	minerals	
igneous	magma	plants	

Rocks are classified into three groups based on the

way they form. A rock that formed from melted rock is

called a(n) _____igneous_____ rock. Rocks with large

mineral grains formed from _____magma_____. Rocks

with small mineral grains formed from _____lava_____.

Shale is a(n) _____sedimentary_____ rock because it formed

when tiny bits of rocks pressed together in layers.

Other rocks of this kind can contain tiny bits of

_____animals_____ and _____plants_____. When rocks

are heated and squeezed inside Earth, new rocks called

_____metamorphic_____ rocks can form. Rocks are made of solid,

nonliving materials called _____minerals_____. They can be

identified by their _____hardness_____, _____luster_____,

and streak. Rocks and minerals are very useful.

Marble Memorials

Read the Writing in Science feature in your textbook.

> ### Write About It
> **Descriptive Writing** Choose two objects made from rock. Write a paragraph that describes and compares them.

Getting Ideas

Write the names of the two objects above the ovals below. In the outer part of each oval, write how they are different. In the overlapping part, write how they are alike.

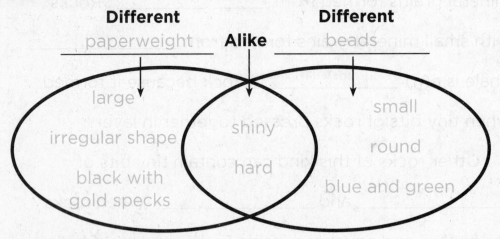

Different
paperweight

Alike

Different
beads

large
irregular shape
black with gold specks

shiny
hard

small
round
blue and green

Planning and Organizing

Lily wrote two sentences. Write "compare" or "contrast" depending on whether each sentence is alike or different.

1. ___compare___ Both necklaces were made of blue stones.

2. ___contrast___ Another necklace had smooth stones, but the stones were black.

© Macmillan/McGraw-Hill

Drafting

Begin your paragraph by writing a sentence that identifies the two objects you will compare. Write a main idea about them.

Possible answer: I have two necklaces made of different stones.

Now write your paragraph. Use a separate piece of paper. Start with the sentence you wrote above. Then compare the two things and include details.

Revising and Proofreading

Here is part of a paragraph that Lily wrote. She made five mistakes. Proofread the sentences. Find the mistakes and correct them.

There are two statues that I like. both of them are made of marble. One statu is made of white marble. It is a sculpture of a jack rabit. The other statue is made of black marble. It is a sculpture of a gient black spider. The marble on both sculptures is very smooth and cold. Even when it's hot outside the marble is still cold.

Now revise and proofread your writing. Ask yourself:

▶ Did I compare two things made from rocks?

▶ Did I use details that show how they are alike and different?

▶ Did I correct all mistakes?

Soil

Use your textbook to help you fill in the blanks.

What is soil?

1. Minerals, weathered rocks, and bits of decayed plants and animals make up _____ soil _____ .

2. Plants use nutrients that _____ humus _____ adds to soil.

3. A plant's _____ roots _____ take in water and hold the soil in place.

4. Bits of rock, minerals, and a lot of humus make up the soil layer called _____ topsoil _____ .

5. The soil layer called _____ subsoil _____ has less humus than the layer of soil above it.

6. Below topsoil and subsoil is solid rock, or _____ bedrock _____ .

How are soils different?

7. Soils with thick layers and lots of humus are good for _____ growing plants _____ .

8. Soils are different _____ colors _____ because they have different minerals and amounts of humus.

9. When you say that soil has large or small grains, you are describing the _____ texture _____ of the soil.

© Macmillan/McGraw-Hill

10. The type of soil with the largest grains is _____sandy soil_____ .

11. Soil with grains smaller than those of sand but larger

than clay is _____silty soil_____ .

12. Some plants may not grow well in _____clay soil_____
because it is too wet.

13. The best soil for growing many plants is _____loam_____ .

Why is soil important?

14. Soil is a(n) _____natural resource_____ and is important
because people need the plants that grow in soil.

15. People can keep soil _____healthy_____ by preventing
soil erosion and keeping soil clean.

Critical Thinking

16. If you were a farmer looking to buy land to grow
plants, which kind of soil would you look for?
Describe the qualities you would like the soil to have.

Possible answer: I would look for soil that has a rich layer of

topsoil, that has plenty of humus, and that is dark brown or

black. Loam would be good because it is not too dry or wet.

_____

_____

_____

© Macmillan/McGraw-Hill

Name _____ Date _____

Soil

Match the correct word to its description below. Write the letter of the word in the blank provided.

a. bedrock	**e.** soil
b. humus	**f.** subsoil
c. loam	**g.** topsoil
d. natural resource	

1. ____c____ soil made up of a mixture of sand, silt, and clay

2. ____g____ the top layer of soil

3. ____f____ the layer of soil that has a lighter color and less humus than the layer above it

4. ____b____ bits of decayed plants and animals that add nutrients to soil

5. ____d____ material on Earth that is necessary or useful to people

6. ____a____ solid rock

7. ____e____ a mixture of minerals, weathered rocks, water, air, and living things

© Macmillan/McGraw-Hill

Soil

Use the words in the box to fill in the blanks below.

clay soil	minerals	subsoil
humus	mixture	topsoil
loam	natural resource	weathering

Soil is important because plants need it to grow.

Because soil is found in nature and is useful to people,

it is a(n) __natural resource__ . Soil is made up of

minerals, weathered rocks, and __humus__ . The

soil in which plants grow well is called __loam__ .

This kind of soil is a(n) __mixture__ of sand, silt,

and clay. Sandy soil holds little water, and __clay soil__

holds a lot of water.

Soils vary because they contain different rocks and

__minerals__ . Soil starts forming when rocks are

broken down by __weathering__ . The first layer of

soil is called __topsoil__ and the next layer is

called __subsoil__ . Solid rock called bedrock is

below the soil layers.

© Macmillan/McGraw-Hill

Fossils and Fuels

Use your textbook to help you fill in the blanks.

How are fossils formed?

1. The bone of an animal that lived long ago is an

   example of a(n) _____*fossil*_____ .

2. An animal footprint in solid rock is a type of fossil

   called a(n) _____*imprint*_____ .

3. Fossils can be in the form of bones, leaves, skin,

   footprints, or _____*shells*_____ .

4. An organism buried in sediment can turn into a type

   of fossil called a(n) _____*stony model*_____ .

5. An empty space in rock in the shape of a living

   thing's remains is a(n) _____*mold*_____ .

6. A copy of a mold's shape that formed in hardened

   minerals is a(n) _____*cast*_____ .

What are fossil fuels?

7. People get energy to heat homes by burning material

   called _____*fuel*_____ .

8. After ancient plants and animals died, their remains

   turned into a fuel called a(n) _____*fossil fuel*_____ .

9. Fossil fuels, plants, animals, water, and air are _____*natural resources*_____ .

© Macmillan/McGraw-Hill

10. Plants and animals are _____renewable_____ resources because they can be replaced.

11. Oil and gas are _____nonrenewable_____ resources because they cannot be replaced.

What are some other sources of energy?

12. We have many _____sources_____ of energy besides fossil fuels.

13. A renewable resource that comes from the Sun is _____solar energy_____ .

14. People can use the Sun, wind, and moving water to make _____electricity_____ .

Critical Thinking

15. Which kind of fossil would you like to discover and what do you think can be learned from it?

Possible answer: I would like to discover a stony model of an

organism from a long time ago so that I can learn about what

the organism looked like and research how it lived.

_____

_____

_____

© Macmillan/McGraw-Hill

Fossils and Fuels

**Match the correct word with its description below.
Write its letter in the blank provided.**

a. cast	**e.** mold
b. fossil	**f.** nonrenewable resource
c. fuel	**g.** renewable resource
d. imprint	**h.** solar energy

1. _____a_____ a type of fossil that is a copy of a mold's shape

2. _____d_____ a type of fossil that is a mark in solid rock

3. _____h_____ energy from the Sun

4. _____g_____ a resource that can be replaced or used again and again

5. _____c_____ a material that is burned for its energy

6. _____e_____ a type of fossil that is an empty space in rock where the remains of an animal or plant lay

7. _____f_____ a resource that cannot be replaced or reused easily

8. _____b_____ the trace or remains of something that lived long ago

© Macmillan/McGraw-Hill

Fossils and Fuels

Use the words in the box to fill in the blanks below.

cast	imprint	nonrenewable	solar energy
energy	minerals	organisms	stony model
fossils	mold	sediment	

Fossil fuels are an important natural resource.

These fuels formed from the remains of _____ organisms

that lived long ago. Remains or traces of ancient plants

and animals are called _____ fossils . A footprint of

an animal in solid rock is a(n) _____ imprint .

Organisms that have died can be buried in _____ sediment .

As the sediment becomes rock, _____ minerals replace

the hard parts of the organism, making a(n) _____ stony model .

A space in the shape of an organism's remains is a(n)

_____ mold . If this fossil fills with water, minerals can

harden and form a(n) _____ cast . Because fossils

take so long to form, they are _____ nonrenewable resources.

However, fossils are not our only source of _____ energy .

Other sources include wind, moving water, and _____ solar energy .

These sources are renewable because they can be replaced.

Turning the Power On

Read the Reading in Science feature in your textbook.

Fill in the blanks in the graphic organizer below.
When you have finished, you will see how text
clues help you draw conclusions.

Text Clues	Conclusions
People need _____ energy _____ . Some energy sources like _____ coal or oil _____ will be used up one day. Some energy sources are _____ renewable _____ and can be used again and again.	People need to use more _____ renewable _____ energy sources so that nonrenewable energy sources will not be used up.
Hydropower, wind, geothermal, solar, and biomass energy can all produce _____ electricity _____ . This form of energy _____ powers _____ our cars, _____ heats _____ our homes, and _____ runs _____ our machines.	Renewable resources can supply our need for _____ electricity or energy _____ .

© Macmillan/McGraw-Hill

Write About It

Draw Conclusions Why is it important for people to use renewable energy sources? Use what you already know and what you read in the article to draw a conclusion.

Planning and Organizing

Answer the following questions.

1. Why do people need energy?

People need energy to power their cars, heat their homes, and

run their machines.

2. What will happen if people use coal and oil instead of renewable resources?

These resources will be used up.

3. Can renewable resources meet people's energy needs? Explain.

Yes. Renewable resources can be replaced and will not be used

up.

Draw Conclusions

Now, use your answers to the questions above to write an answer to this question: "Why is it important for people to use renewable resources?"

Possible answer: It is important for people to use renewable

resources because they can be replaced, so people's energy needs

will always be met.

© Macmillan/McGraw-Hill

Air and Water Resources

Use your textbook to help you fill in the blanks.

How do we use air and water?

1. Air is important because it has _____oxygen_____ that people need in order to breathe.

2. Water is a(n) _____renewable resource_____ because it can be replaced by rain or snow.

3. People cannot drink most of Earth's water because it is very _____salty_____ .

4. Rivers, ponds, and water below ground, or _____groundwater_____ , are fresh-water sources.

5. People cannot use most of Earth's fresh-water sources because the water is _____frozen_____ .

How do people get water?

6. Water is carried to people through _____aqueducts_____ .

7. A wall that is built on a river is called a(n) _____dam_____ .

8. People can use the water that is stored in a(n) _____reservoir_____ behind a dam.

9. People dig _____wells_____ to pump groundwater to Earth's surface.

10. At a water treatment plant, water is cleaned so that it is _____safe_____ to drink.

What can happen to air and water resources?

11. Living things can become sick from _____pollution_____
    to water, air, or land.

12. Air becomes polluted when people burn _____fossil fuels_____ .

13. Water becomes polluted when people use _____fertilizers_____
    to make plants grow.

14. If people _____waste_____ water, they may
    eventually use all of the available water on Earth.

How can you conserve resources?

15. When people use less water, they _____conserve_____
    resources.

Critical Thinking

16. Why are Earth's air and water resources important?
    How might these resources be harmed by human
    activities?

    Possible answer: Air is important because people need the

    oxygen in air to breathe. Water is important because people

    need it to farm, cook, make things, and swim. If air or water get

    polluted, living things can become sick. If people waste water, it

    can be used up before it is replaced.

    _____

    _____

Name _____ Date _____

Air and Water Resources

Match each word with its description. Then write its letter in the blank provided.

a. aqueducts	**c.** groundwater	**e.** reservoir
b. conserve	**d.** pollution	**f.** well

1. ___f___ a hole that has been dug to reach underground water

2. ___b___ to use resources wisely

3. ___a___ pipes or ditches that carry water

4. ___e___ water stored behind a dam

5. ___d___ harmful things in the water, air, and land

6. ___c___ water held in rocks below ground

Air and Water Resources

Use the words in the box to fill in the blanks below.

air	fresh water	pollution	water treatment
aqueducts	groundwater	renewable	wells
fossil fuels	nature	reservoirs	

We must use resources wisely so that we will not run out of them. Two important resources are water and _____air_____ . These resources are _____renewable_____ , but they can be harmed by _____pollution_____ . Volcanoes and wildfires are pollution caused by _____nature_____ . People cause air pollution when they burn _____fossil fuels_____ . Fertilizers can pollute rivers, lakes, and _____groundwater_____ , an important source of _____fresh water_____ .

People get groundwater by digging _____wells_____ . Some people do not live near fresh water. They use _____reservoirs_____ to store water and _____aqueducts_____ to transport water to them. Some cities and towns use _____water treatment_____ plants to make water clean and safe. Because fresh water is limited, people should practice conservation.

Using Earth's Resources

Circle the letter of the best answer.

1. Which of the following is a solid, nonliving substance found in nature?

   a. sediment

   b. humus

   c. a mineral

   d. a fossil

2. Which of the following is made up of bits of decayed plants and animals that add nutrients to soil?

   a. fossils

   b. topsoil

   c. loam

   d. humus

3. Which of the following is an example of a nonrenewable resource?

   a. air

   b. soil

   c. coal

   d. water

4. Which kind of rock forms from magma?

   a. metamorphic

   b. igneous

   c. sedimentary

   d. fossil

5. Which of the following is a mixture of minerals, weathered rocks, and living things?

   a. soil

   b. humus

   c. a fossil

   d. nutrients

6. A dinosaur's footprint in hardened mud is a(n)

   a. mold.

   b. cast.

   c. stony model.

   d. imprint.

© Macmillan/McGraw-Hill

Circle the letter of the best answer.

7. What do people use to get water from below Earth's surface?

   a. aqueducts

   b. wells

   c. reservoirs

   d. dams

8. Rock formed from heat and pressure inside Earth is classified as

   a. bedrock.

   b. igneous rock.

   c. sedimentary rock.

   d. metamorphic rock.

9. Which of the following is Earth's main source of energy?

   a. fossil fuels

   b. magma

   c. the Sun

   d. moving water

10. What kind of resource are plants, animals, water, and air?

    a. nonrenewable

    b. renewable

    c. energy

    d. limited

11. Which kind of rock forms from tiny bits of plants, animals, or weathered rock?

    a. sedimentary

    b. metamorphic

    c. igneous

    d. mineral

12. Which of the following is a material that is burned for its energy?

    a. magma

    b. solar energy

    c. fossils

    d. fuel

What a Difference Day Length Makes

Read the Unit Literature feature in your textbook.

Write About It

Response to Literature Animals respond to changing seasons in many ways. What are some ways you have seen nature change from season to season? Write about it.

Paragraphs should have a clear topic sentence that directly

addresses ways that students have seen nature change from

season to season. The sentences that follow the topic sentence

should support the topic sentence by providing details, such as

the changing of color of leaves in autumn. Students should use a

closing sentence that restates the main idea of the paragraph. Good

paragraphs should include correct grammar and mechanics, and

demonstrate a proper transition from one idea to another.

_____

_____

_____

_____

Changes in Weather

Complete the concept map with information that you learned about the events in the water cycle. Some information has been written for you.

1. The Sun

_____ heats _____

water, causing it to

_____ evaporate _____

from Earth's lakes, rivers, and oceans.

The Water Cycle

2. When water evaporates, it forms tiny drops called

_____ water vapor _____,

which cannot be seen.

5. Precipitation, such as rain,

_____ snow _____,

or sleet, falls into Earth's

_____ oceans, rivers, and _____

_____ lakes _____.

3. Water vapor

_____ rises _____

into the

_____ atmosphere _____.

4. Clouds form when water vapor

_____ condenses _____

around tiny

_____ dust _____

particles in the air.

© Macmillan/McGraw-Hill

Name _____ Date _____

Weather

Use your textbook to help you fill in the blanks below.

What is weather?

1. Weather is what the _____air_____ is like in the lowest layer of the atmosphere at a certain time and place.

2. Air is a(n) _____gas_____ that takes up space and can move things.

3. The air that surrounds Earth makes up part of the _____atmosphere_____ .

4. Earth's atmosphere consists of layers of _____gases_____ and some _____dust_____ .

5. The measure of how hot or cold something is can be found by taking its _____temperature_____ .

How can you describe the weather?

6. Weather is described in terms of amounts of _____precipitation_____ and _____wind_____ , as well as _____air pressure_____ and air temperature.

7. Water that falls from the _____atmosphere_____ to the ground is called precipitation.

© Macmillan/McGraw-Hill

8. Two forms of precipitation are _____rain_____

and _____snow_____ .

9. Air that moves is called _____wind_____ .

10. The weight of air pressing down on Earth is

called _____air pressure_____ .

How do we predict weather?

11. Scientists collect data about the atmosphere by

using _____weather balloons_____ .

12. To observe weather from above Earth, _____satellites_____
are used.

13. Data from _____maps_____ , balloons, and satellites

are used to _____predict_____ weather.

Critical Thinking

14. What information would you need to collect to
describe the weather in your town every day
for a week?

Possible answer: I would need to know the temperature every

day. I would also need to know the air pressure, the wind speed,

and the wind direction. I would have to note any precipitation

and record it as rain, hail, sleet, or snow.

_____

Weather

Match each word from the box to its definition.

a. atmosphere	**c.** precipitation	**e.** temperature	**g.** weather
b. hail	**d.** sleet	**f.** thermometer	**h.** wind

1. _____d_____ a type of precipitation that involves mixed states of matter

2. _____c_____ water that falls to the ground from the atmosphere

3. _____e_____ the measure of how hot or cold something is

4. _____g_____ what air is like at a certain time and place

5. _____a_____ the layers of gases and tiny bits of dust that surround Earth

6. _____b_____ lumps of ice that fall during a thunderstorm

7. _____f_____ a tool that measures temperature

8. _____h_____ moving air

Weather

Use the words in the box to fill in the blanks below.

air pressure	moves	speed	weight
atmosphere	precipitation	temperature	wind
direction	rain	thermometer	

Weather forms in the lowest layer of gases that surround Earth. This layer makes up part of Earth's ___atmosphere___ . Scientists use a tool called a(n) ___thermometer___ to measure the ___temperature___ of air. Air takes up space and has ___weight___ . The weight of air as it presses down on Earth is called ___air pressure___ .

Air also ___moves___ . This movement is called ___wind___ . There are tools that measure the ___direction___ and ___speed___ of wind. Occasionally, ___precipitation___ falls from the atmosphere to Earth. Precipitation may be in the form of sleet, snow, or ___rain___ . These are all different forms of water.

The Water Cycle

Use your textbook to help you fill in the blanks below.

What are clouds?

1. A collection of tiny drops of water or ice that can be

   seen in the air is a(n) _____cloud_____ .

2. Low, flat layers of clouds covering most of the sky

   are _____stratus_____ clouds.

3. On a sunny day, it is possible to see white, puffy

   clouds with flat bottoms, or _____cumulus_____ clouds.

How do clouds form?

4. A stratus cloud that forms near the ground is _____fog_____ .

5. Liquid water _____evaporates_____ and becomes a(n)

   _____gas_____ when the Sun shines on it.

What is the water cycle?

6. The path that water takes between Earth's

   surface and the _____atmosphere_____ is called the
   water cycle.

7. The Sun causes the _____evaporation_____ of water by
   heating it.

© Macmillan/McGraw-Hill

8. Clouds form as water condenses on ___tiny specks of dust___

   in the air and returns to Earth as ___precipitation___ ,
   continuing the cycle.

What are some kinds of severe weather?

9. A storm that forms over an ocean and brings heavy

   rain and strong winds is a(n) ___hurricane___ .

10. A powerful storm shaped like a tall funnel that forms

    over land is a(n) ___tornado___ .

11. A blizzard has strong winds, snow, and ___cold temperatures___ .

How can you stay safe in severe weather?

12. During most severe storms, ___stay inside___

    a building. During a tornado, go to a(n) ___basement___

    or lie flat in a(n) ___low place___ .

Critical Thinking

13. How are clouds related to the water cycle
    and storms?

    Clouds form as part of the water cycle. Heat from the Sun

    causes water to evaporate from Earth's surface. Water vapor

    rises and condenses around pieces of dust, forming clouds. The

    water that falls from clouds may be part of a thunderstorm,

    blizzard, or hurricane.

The Water Cycle

What am I?

Choose a word from the box below that answers each question. Write its letter in the space provided.

a. cloud	**d.** hurricane	**g.** water cycle
b. cumulus clouds	**e.** stratus clouds	
c. evaporation	**f.** tornado	

1. I am white and puffy with a flat bottom. What am I? _____b_____

2. I form over the ocean and come with strong winds and heavy rain. What am I? _____d_____

3. I form when water vapor rises and cools around tiny dust particles. What am I? _____a_____

4. I have a long funnel and cause damage on land. What am I? _____f_____

5. I am made of low, flat layers of clouds. What am I? _____e_____

6. I am the path that water takes from Earth to the atmosphere and back again. What am I? _____g_____

7. I am the process by which a liquid changes to a gas. What am I? _____c_____

The Water Cycle

Use the words in the box to fill in the blanks below.

clouds	precipitation	water cycle
dust	Sun	water vapor
evaporation	tornado	

Every day, water moves from Earth to the

atmosphere and back again. This process is called the

_____water cycle_____ . During the water cycle, the

_____Sun_____ heats water on Earth's surface. This

causes the _____evaporation_____ of water, changing it to

a gas. The _____water vapor_____ rises and condenses

around _____dust_____ particles in the atmosphere,

forming _____clouds_____ . Water that falls from

clouds is _____precipitation_____ .

Hurricanes, thunderstorms, tornadoes, and

blizzards are all types of severe weather. During a(n)

_____tornado_____ , it is best to find shelter in a

basement or to lie flat in a low area. People must be

aware of the dangers of extreme weather.

Name _____ Date _____

Tracking Twisters

Read the text below, and answer the questions that follow.

When a tornado, or twister, touches down, it can destroy almost anything in its path. For this reason, scientists gather information about tornadoes to help predict where they may happen.

First, scientists observe and measure weather to see if conditions are right for a tornado to form. Tornadoes occur when warm, moist air near the ground mixes with cool, dry air above it and rises rapidly.

Doppler radar is used to track storms. Radar works by sending out radio waves from an antenna. Objects in the air, such as raindrops, bounce the waves back to the antenna. Doppler radar can track the direction and speed of a moving object, such as a tornado or other storm.

People called storm chasers get a close-up look at tornadoes from planes or cars. The information they gather is used to warn communities about tornadoes before they strike.

1. What might happen if a tornado touches down?

It could destroy anything in its path.

2. What is Doppler radar used to track?

It is used to track storms.

3. How do storm chasers help communities?

They warn them about tornadoes before they strike.

Write About It

Predict What if there were no storm chasers? What if there was no technology to warn people of tornadoes? Write about what might happen.

Fill in the graphic organizer below.

What I Predict	What Happens
A(n) _____tornado_____ touches down.	It may _____destroy_____ anything in its path.
Scientists do not have _____Doppler radar_____ to track storms.	They cannot track the _____direction_____ and _____speed_____ of a tornado.
There were no _____storm chasers_____ to gather _____information_____ .	Communities could not be _____warned_____ about tornadoes before they _____strike_____ .

Write a paragraph telling what might happen if there were no storm chasers or technology to warn people about tornadoes.

Answers will vary but should demonstrate an understanding that

people cannot prepare for a tornado strike if they are not warned

ahead of time. Answers might include examples of safety measures

that people would not take, such as staying outside instead of going

to a basement.

Climate and Seasons

Use your textbook to help you fill in the blanks below.

What is climate?

1. Every day, the _____weather_____ changes.

2. The pattern of weather in a certain place over a long

   time is its _____climate_____ .

3. The climate of a certain place is described by

   its average _____temperature_____ and amount of _____precipitation_____ .

4. Earth is shaped like a ball, or a(n) _____sphere_____ .

5. Earth spins around its _____axis_____ , a(n) _____line_____
   through the center of a spinning object.

6. Earth's axis is _____tilted_____ slightly.

7. Places on Earth with _____warm_____ climates are
   hit directly by the Sun's rays.

8. When the Sun's rays strike Earth at a(n) _____slant_____ ,
   energy is spread out more, so the areas that receive
   these rays have colder climates.

What affects climate?

9. The climate becomes _____colder_____ in areas that
   are high in the atmosphere.

10. Mountains next to an ocean can block moist air from

going inland and force that air _____upward_____ ,
causing precipitation.

11. On the ocean side of a mountain, the climate is often

_____wet_____ , and on its opposite side, the

climate is often _____dry_____ .

What are seasons?

12. Times of year with different weather patterns

are _____seasons_____ .

13. The four seasons are _____summer_____ , _____fall_____ ,
winter, and spring.

Critical Thinking

14. How are climate and weather different?

Weather is what the air is like at a certain time and place.

Weather changes from day to day. Climate is the pattern of

weather in a certain place over a long time. Climate does not

change every day.

_____

_____

_____

Climate and Seasons

Match the word with its definition.

a. axis	**c.** mountain	**e.** seasons	**g.** sphere
b. climate	**d.** ocean	**f.** slant	**h.** weather

1. ___f___ the direction in which the Sun's rays strike Earth, causing energy to be spread out

2. ___b___ the pattern of weather in a certain place over a long time

3. ___g___ the shape of Earth

4. ___e___ times of the year with different weather patterns

5. ___d___ a landform that keeps the air temperature of nearby land from becoming too hot or cold, producing a milder climate

6. ___a___ the real or imaginary line through the center of a spinning object

7. ___c___ affects how wet a climate is by blocking moist air from some places

8. ___h___ changes every day

Climate and Seasons

Use the words in the box to fill in the blanks below.

axis	milder	slant
climate	ocean	temperatures
colder	precipitation	warmer

Weather changes every day. The _____climate_____

stays the same. The yearly ____temperatures____ and

____precipitation____ of an area describe the climate.

The Sun also affects climate. Earth spins on a slightly

tilted _____axis_____ . Because of this, the Sun's

rays strike some places on Earth directly. These places

get much more of the Sun's energy and have

_____warmer_____ climates. In other places, the Sun's

rays strike an area on a(n) _____slant_____ . These

areas get less of the Sun's energy, so the climate is

_____colder_____ . Land that is near a(n) ____ocean____

or large lake has a(n) _____milder_____ climate. Land

on the far side of mountains near an ocean tends

to be dry.

© Macmillan/McGraw-Hill

Name _____ Date _____

A Season Myth

Read the Writing in Science feature in your textbook.

> ### Write About It
>
> **Fictional Narrative** Write your own myth about how a season came to be. Use an animal as a character in the story. Use all the parts of a good story when writing.

Getting Ideas

Choose a season. How could you create a myth about how this season came about? Fill in the sequence chart below. Your main character should be an animal.

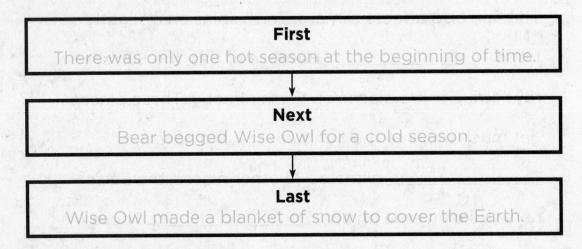

| **First** |
| There was only one hot season at the beginning of time. |

↓

| **Next** |
| Bear begged Wise Owl for a cold season. |

↓

| **Last** |
| Wise Owl made a blanket of snow to cover the Earth. |

Planning and Organizing

Jake wrote three sentences. Put the sentences in time order. Write 1, 2, or 3 in each blank.

1. ___3___ Wise Owl agreed to cover the land with snow.

2. ___2___ Bear wanted a cold season, so he visited Wise Owl.

3. ___1___ Long ago, there was only one hot season.

Drafting

Begin your myth. Write a sentence that introduces the main character. Describe what it was like in the beginning of time. Get the reader interested!

Possible answer: Big Bear shook his furry coat and said, "I'm tired of

being hot all the time."

Now write your story on a separate piece of paper. Start with the sentence you wrote above. Tell about the character and where and when the story takes place.

Revising and Proofreading

Jake wrote some sentences and wants to join them together. Read the sentences and look at the two words that follow. Circle the word that best connects the two sentences. Then write the new sentence on the line, adding a comma before the connecting word.

1. The Earth was snowy. All of the animals slept or hid in trees. so but

   The Earth was snowy, so all of the animals slept or hid in trees.

2. Bluebird wanted to sing. It was too cold. and but

   Bluebird wanted to sing, but it was too cold.

Now revise and proofread your writing. Ask yourself:

▶ Did I tell how a season came to be?

▶ Did I include a beginning, middle, and end?

▶ Did I correct all mistakes?

Name _____ Date _____

Changes in Weather

Circle the letter of the best answer.

1. The warmest season is

   a. winter.

   b. spring.

   c. summer.

   d. fall.

2. Rain, snow, and sleet are examples of

   a. clouds.

   b. condensation.

   c. evaporation.

   d. precipitation.

3. During condensation, a

   a. solid changes to a liquid.

   b. liquid changes to a solid.

   c. gas changes to a liquid.

   d. solid changes to a gas.

4. Earth's air, land, and water are heated by

   a. clouds.

   b. precipitation.

   c. condensation.

   d. the Sun's energy.

5. In the water cycle, the Sun's energy heats water and causes

   a. condensation.

   b. evaporation.

   c. precipitation.

   d. boiling.

6. A person can stay safe during a lightning storm by

   a. standing under a tree.

   b. using appliances.

   c. lying flat in a low place.

   d. staying inside a building.

7. A thermometer is used to measure

   a. air pressure.

   b. the weight of air pressing on Earth.

   c. air temperature.

   d. wind speed and direction.

Circle the letter of the best answer.

8. Which type of cloud is thin and wispy and forms high above the ground?

   a. fog

   b. cirrus

   c. stratus

   d. cumulus

9. Which type of large, severe storm forms strong winds and heavy rains over the ocean?

   a. tornado

   b. hurricane

   c. blizzard

   d. thunderstorm

10. Places on the ocean side of mountains tend to have a

   a. hot climate.

   b. wet climate.

   c. dry climate.

   d. cold climate.

11. The layers of gases and dust that surround Earth make up the

   a. atmosphere.

   b. Sun's rays.

   c. four seasons.

   d. fog.

12. Parts of Earth that are hit directly by the Sun's rays have

   a. warmer climates.

   b. colder climates.

   c. mild climates.

   d. summer all year long.

13. Around which of the following items does water vapor condense to make clouds in the atmosphere?

   a. fog on the ground

   b. specks of dust

   c. hail

   d. sleet

Planets, Moons, and Stars

Complete the concept map with causes and effects
described in the chapter.

Cause →	Effect
Earth _____rotates_____ on its _____axis_____.	Earth has both day and _____night_____.
Earth _____revolves_____ around the Sun.	Earth has seasons.
Earth's _____axis_____ is tilted.	Earth has opposite _____seasons_____ in the Northern and Southern Hemispheres at any given time.
The _____Moon_____ revolves around _____Earth_____.	We see phases of the Moon.
Sunlight _____reflects_____ off of the Moon and planets.	The Moon and planets appear to shine.
The Sun is the _____closest_____ star to Earth.	The Sun appears to be _____biggest_____ in size.
Stars are big balls of _____hot gases_____.	Stars give off _____heat_____ and _____light_____.

© Macmillan/McGraw-Hill

The Sun and Earth

Use your textbook to help you fill in the blanks.

Why is there day and night?

1. The Sun seems to move across the sky because Earth spins, or _____rotates_____ .

2. The side of Earth that faces the Sun has _____daytime_____ .

3. The Sun appears to rise in the _____eastern_____ sky and is highest in the sky at _____midday_____ .

4. Shadows are long at _____sunrise_____ and at _____sunset_____ .

5. The imaginary line on which Earth spins is its _____axis_____ .

6. Earth turns once on its axis every _____24_____ hours.

Why are there seasons?

7. The path that Earth follows as it moves around the Sun is called a(n) _____orbit_____ .

8. It takes 365 days, or one _____year_____ , for Earth to revolve around the Sun.

9. Seasons change because Earth's axis is _____tilted_____ and different areas of Earth are closer or farther away from the Sun.

10. When the Northern Hemisphere is tilted toward the

Sun, it has _____summer_____ .

11. When it is summer in the Northern Hemisphere, it is

_____winter_____ in the Southern Hemisphere.

12. During winter, days are _____shorter_____ than they
are in summer.

What is the Sun like?

13. Earth revolves around a(n) _____star_____ called
the Sun.

14. Life on Earth exists because the Sun provides

_____light_____ and _____heat_____ .

Critical Thinking

15. If summer starts on June 21 in the Northern
Hemisphere, when does summer start in the Southern
Hemisphere? Explain.

Summer starts in the Southern Hemisphere on December 21.

When the Southern Hemisphere is tilted toward the Sun it is

summer there. At that time, the Northern Hemisphere is tilted

away from the Sun and having winter.

_____

_____

The Sun and Earth

**Match the words in the box to their definitions below.
Then find the words hidden in the puzzle.**

a. axis	**c.** orbit	**e.** rotation	**g.** year
b. day	**d.** revolve	**f.** star	

1. _____a_____ a real or imaginary line through the center of a
spinning object

2. _____d_____ to move around another object

3. _____c_____ a regular path Earth follows around the Sun

4. _____g_____ the time it takes Earth to make one complete trip
around the Sun

5. _____f_____ a ball of hot, glowing gases

6. _____b_____ the time it takes Earth to make one complete turn

7. _____e_____ the movement of Earth when it spins

R	O	T	A	T	I	O	N
X	A	M	D	S	B	H	W
S	Y	R	V	T	M	T	D
A	E	J	I	A	I	O	A
P	A	B	Q	R	D	F	Y
L	R	E	V	O	L	V	E
O	N	A	X	I	S	V	P

Name _____ Date _____

The Sun and Earth

Use the words in the box to fill in the blanks below.

axis	orbit	summer	winter
daytime	rotating	tilted	year

Earth's movement around the Sun causes daytime,

nighttime, and seasons. Earth follows a(n) ___orbit___

around the Sun. A complete trip takes one ___year___ .

As Earth revolves around the Sun, different parts of

Earth are ___tilted___ toward or away from

the Sun. If the upper half of Earth is tilted toward the

Sun, it is ___summer___ in the Northern

Hemisphere. At the same time, the lower half of Earth

is tilted away from the Sun, and it is ___winter___

in the Southern Hemisphere.

Earth also moves by ___rotating___ on its

___axis___ every 24 hours. When one side of

Earth faces the Sun, it is ___daytime___ . On the

side that faces away from the Sun, it is nighttime.

Seasons Where You Live

Write About It

Personal Narrative Choose a season. Tell a true story about something you did during that season. Explain why you still remember the event. How did it make you feel? Describe what the weather was like.

Getting Ideas

Think about something you did this season. Write the events in the chart below. Write them in time order.

Possible answer:

Snow fell.

↓

I put on snowshoes.

↓

I hiked to the cabin in the woods.

Planning and Organizing

Here are three sentences that Brian wrote. Put them in time order. Write 1 by the first event, 2 by the second event, and 3 by the last event.

____1____ The snow fell heavily all through the night.

____2____ The next morning, I strapped on my snowshoes and set out.

____3____ I hiked through the woods to the cabin for steaming hot cider.

Drafting

Write a sentence to begin your personal narrative. Use "I" to refer to yourself. Include the name of the season you are writing about.

Possible sentence: Last winter, I walked through the woods.

Now write your personal narrative. Use a separate piece of paper. Begin with the sentence you wrote. Write about what you did in time order. Remember to describe the weather. Explain why this event is memorable.

Revising and Proofreading

Here are some sentences that Brian wrote. He did not use the first-person pronoun "I" to refer to himself. Rewrite each sentence by changing the pronouns to the first person.

1. He looked up in wonder at the branches covered with snow.

   I looked up in wonder at the branches covered with snow.

2. Snow fell on his head when two squirrels jumped on a branch.

   Snow fell on my head when two squirrels jumped on a branch.

3. He laughed as he shook the snow off his cap.

   I laughed as I shook the snow off my cap.

Now revise and proofread your writing. Ask yourself:

▶ Did I tell about something that happened to me?

▶ Did I use the pronoun "I"?

▶ Did I correct all mistakes?

The Moon and Earth

Use your textbook to help you fill in the blanks.

What are the phases of the Moon?

1. The different shapes that we see of the Moon are

   called _____ phases _____ .

2. There are _____ eight _____ main phases of
   the Moon.

3. The Moon seems to move across the sky because of

   Earth's _____ rotations _____ .

4. The Moon seems to be lit because _____ sunlight _____
   reflects off of it.

5. Because the Moon revolves around Earth, it is

   a(n) _____ satellite _____ of Earth.

Why does the Moon's shape seem to change?

6. The Moon is always a(n) _____ sphere _____ .

7. The Sun always lights one _____ half _____ of the
   Moon's surface.

8. When we see a phase of the Moon, we see some of

   the Moon's _____ lighted _____ half.

9. The dark half of the Moon faces Earth during the

   _____ new-Moon _____ phase.

10. The dark half of the Moon faces away from Earth

during the _____full-Moon_____ phase.

11. One complete cycle of the Moon's phases takes

about _____29 days_____ .

What is it like on the Moon?

12. The Moon and Earth both get _____light_____ from the Sun.

13. Three features that Earth has but the Moon does not

have are _____liquid water_____ , a(n) _____atmosphere_____ ,

and _____living things_____ .

14. Unlike Earth, the Moon is covered with _____craters_____ .

15. Craters are made when _____rocks_____ crash into the surface of the Moon.

Critical Thinking

16. A new Moon can be seen when the Moon is between Earth and the Sun. How do these three objects line up when we see a full Moon? How do you know?

During a full Moon, Earth is situated between the Sun and the

Moon. The entire lighted half of the Moon is facing Earth, so the

Sun must be on the other side of Earth, opposite the Moon.

_____

© Macmillan/McGraw-Hill

The Moon and Earth

Match the correct word with its definition. Write its letter in the space provided.

a. crater	**d.** full Moon	**g.** new Moon
b. crescent Moon	**e.** gibbous Moon	**h.** phases
c. first-quarter Moon	**f.** last-quarter Moon	

1. __e__ the phase of the Moon that happens just before and after a full Moon

2. __a__ a hollow area in the ground

3. __g__ the phase of the Moon that happens when the lighted half faces away from Earth

4. __d__ the phase of the Moon that happens when we see the entire lighted half

5. __c__ the phase of the Moon that happens about 7 days after the new Moon

6. __f__ the phase of the Moon that happens about 21 days after the new Moon

7. __b__ the phase of the Moon that happens just before and after a new Moon

8. __h__ the different shapes that we see of the Moon

The Moon and Earth

Use the words in the box to fill in the blanks below.

craters	living things	reflects	rotate
faces	phases	revolve	satellite

In space, the Moon is the closest natural object to

Earth. It is called a(n) _____satellite_____ because it

revolves around Earth. The Moon is smaller than Earth

and has many _____craters_____ . There are no

_____living things_____ on the Moon because it has no

atmosphere or liquid water.

Earth and the Moon _____rotate_____ on their

own axes and _____revolve_____ around larger objects.

As the Moon revolves, we see different amounts of its

lighted side, or _____phases_____ . Half of the Moon is

always lit because sunlight _____reflects_____ off of it.

We see a full Moon when the entire lighted side

_____faces_____ Earth. A new Moon is visible when

the entire lighted side faces away from Earth.

The Planets

Use your textbook to help you fill in the blanks.

What is our solar system?

1. The Sun and the objects that move around it make
up a(n) _____solar system_____ .

2. Large bodies of rock or gas that revolve around a
star are called _____planets_____ .

3. There are _____eight_____ planets that revolve
around the Sun.

4. All of the planets follow a(n) _____orbit_____ as
they _____revolve_____ around the Sun.

5. The planet with the shortest orbit is _____Mercury_____ .

6. The planet with the longest orbit is _____Neptune_____ .

7. Planets look like _____stars_____ when viewed from
Earth because sunlight reflects off of them.

8. Planets look _____smaller_____ than the Moon
when viewed from Earth because they are farther
away.

What are the inner and outer planets?

9. The four planets closest to the Sun are called
the _____inner planets_____ .

10. The inner planets are made of ___rocky material___ and

are some of the ___smallest___ planets in our
solar system.

11. The names of the inner planets are ___Mercury___ ,

Venus, ___Earth___ , and Mars.

12. The four planets farthest from the Sun are called

the ___outer planets___ .

13. The names of the outer planets are Jupiter,

___Saturn___ , ___Uranus___ , and Neptune.

14. The largest planet in our solar system is ___Jupiter___ .

How can we view the planets?

15. To see a planet's surface, we can use a tool called

a(n) ___telescope___ .

16. Telescopes have ___mirrors___ and ___lenses___
that gather light.

Critical Thinking

17. Which planets were discovered with telescopes?
Why?

Uranus and Neptune were discovered with telescopes. We can

see Mercury, Venus, Mars, Jupiter, and Saturn in the sky with our

eyes, but not Uranus and Neptune.

Name _____ Date _____

The Planets

Use the words in the box to fill in the blanks below.

inner planets	outer planets	space probe	telescope
lens	planet	solar system	year

1. A large body of rock or gas that revolves around a

   star is called a(n) _____ planet _____ .

2. The four planets closest to the Sun are called

   the _____ inner planets _____ .

3. A star and the objects that move around it make up

   a(n) _____ solar system _____ .

4. A machine that leaves Earth and travels through

   space is a(n) _____ space probe _____ .

5. The four planets farthest from the Sun are called

   the _____ outer planets _____ .

6. It takes Earth one _____ year _____ to complete a
   trip around the Sun.

7. A clear material that helps us see objects in more

   detail is a(n) _____ lens _____ .

8. A tool used to make faraway objects appear larger is

   a(n) _____ telescope _____ .

The Planets

Use the words in the box to fill in the blanks below.

closest	light	solar system	warmer
farthest	Neptune	stars	
inner planets	outer planets	telescope	

Earth and seven other planets revolve around the

Sun. They form part of a(n) ___solar system___ . The

planets look like ___stars___ because they

appear to shine. However, planets do not make their

own ___light___ but reflect the Sun's light.

Earth is one of the four ___inner planets___ . This

group of planets is ___closest___ to the Sun. This

makes them ___warmer___ than the other planets.

The other planets are called the ___outer planets___ .

This group of planets is ___farthest___ from the

Sun. The planet that is farthest away is ___Neptune___ .

To see the planets beyond Saturn, we must use a(n)

___telescope___ . A tool called a space probe lands

on planets and takes pictures of them.

© Macmillan/McGraw-Hill

The Stars

Use your textbook to help you fill in the blanks.

What are stars?

1. Stars are balls of hot gasses that give off _____light_____

   and _____heat_____ .

2. The only star in our solar system is _____the Sun_____ .

3. The Sun's light keeps us from seeing other _____stars_____

   during the _____day_____ .

4. Because it is closest to Earth, the Sun seems _____bigger_____
   than other stars.

5. Stars differ in size and whether they are _____hotter_____

   or _____brighter_____ than other stars.

6. You can tell how hot a star is by its _____color_____ .

7. The hottest stars are the color _____blue_____ .

8. The coolest stars are the color _____red_____ .

9. Long ago, people thought groups of stars formed

   pictures, which today we call _____constellations_____ .

**Why do we see different stars during
different seasons?**

10. We do not see the same stars each _____season_____ .

11. We see different stars as Earth _____revolves_____ around the Sun and we observe different views of the sky.

12. We see different stars when Earth faces different _____directions_____ .

13. A constellation that we see only in winter is _____Orion_____ .

14. In winter, Orion is in the sky at _____night_____ because we look out into space and away from the Sun.

15. In summer, Orion is in the sky during the _____day_____ because we look out into space in the same direction as the Sun.

Critical Thinking

16. Why might some stars seem brighter in the night sky than other stars?

The stars that seem brighter could be closer to Earth than other

stars. They also could be hotter and brighter than other stars.

_____

_____

_____

_____

_____

© Macmillan/McGraw-Hill

The Stars

Match the correct word to its definition.

a. blue star	**c.** constellation	**e.** red star	**g.** Sun
b. color	**d.** Orion	**f.** star	

1. ____c____ a group of stars that seem to form a picture

2. ____a____ the hottest kind of star

3. ____d____ a constellation that we can see in winter but not in summer

4. ____b____ a way to tell how hot a star is

5. ____f____ a glowing ball of hot gases that gives off light and heat

6. ____e____ the coolest kind of star

7. ____g____ the only star in our solar system

The Stars

Use the words in the box to fill in the blanks below.

closer	different	season
color	gases	Sun
constellations	revolves	sunlight

There are millions of stars in space. We only see stars

at night when _____sunlight_____ does not block them.

We can see pictures called _____constellations_____ that are

formed by groups of stars. One of these is Orion.

Whether we see Orion depends on the _____season_____ .

Orion stays in the same place, but we see _____different_____

parts of the sky as Earth _____revolves_____ around

the Sun.

Stars are balls of hot _____gases_____ . Our

_____Sun_____ is a star, and it is _____closer_____

to Earth than to any other star. Stars differ in _____color_____ ,

size, and brightness. A blue star is the hottest star,

and a red star is the coolest star.

© Macmillan/McGraw-Hill

Meet Orsola De Marco

Do you ever wonder about the stars? Orsola De Marco does. She is an astrophysicist at the American Museum of Natural History in New York. An astrophysicist is a scientist who studies stars. Orsola studies stars that are found together in pairs. As far as we know, our Sun is a star that stands alone. Most stars in the universe have a partner. They are called binary stars.

These binary stars orbit each other at a very close distance. Scientists think that one star is being absorbed by the other. The butterfly-shaped wings are probably caused by gases from the surface of the central star.

Of course Orsola cannot go to the stars to learn about them. Instead, she travels to Arizona, Hawaii, and Chile to use large telescopes. She gazes billions of miles into space to get a good look at binary stars. She watches how the stars affect each other. When a star gets old it becomes larger. If there is another star nearby, it might get eaten up, or absorbed, by the old star. No one is sure what will happen after that. Orsola is working to find out.

Write About It

Summarize Read the article with a partner. List the most important information in a chart. Then use the chart to help you summarize the article. Remember to start with a main-idea sentence and to keep your summary brief.

Planning and Organizing

▶ List the most important information in the article in the chart below.

Most Important Information

Drafting

▶ Start by writing a clear statement that describes the main idea of the article.

▶ Write three supporting details.

▶ Read what you have written. Cross out anything that does not directly support the main idea.

▶ Exchange papers with your partner and ask him or her to check your choice of a main idea. Have your partner also check your choice of supporting details.

Summarize Write your summary on a separate piece of paper. Use your own words. Include the main ideas and details you wrote.

Planets, Moons, and Stars

Circle the letter of the best answer.

1. We have daytime and
   nighttime because Earth

   a. tilts.

   b. orbits.

   c. rotates.

   d. revolves.

2. What is the name for
   the different shapes of
   the Moon?

   a. revolutions

   b. spheres

   c. craters

   d. phases

3. What is the imaginary line
   on which Earth spins?

   a. rotation

   b. revolution

   c. axis

   d. orbit

4. Which of the following best
   describes Jupiter?

   a. inner planet

   b. planet

   c. moon

   d. star

5. Which of the following
   describes the movement of
   an object around another
   object?

   a. phase

   b. tilting

   c. rotating

   d. revolving

6. The Sun and planets are
   part of a(n)

   a. solar system.

   b. constellation.

   c. revolution.

   d. orbit.

Circle the letter of the best answer.

7. Which of the following is a group of stars?

    a. solar system

    b. constellation

    c. satellite

    d. hot gases

8. Which of the following does Earth complete in one year?

    a. phase

    b. rotation

    c. season

    d. orbit

9. Which of the following takes pictures of planets?

    a. space probe

    b. telescope

    c. lens

    d. constellation

10. Which of the following gives off light?

    a. star

    b. moon

    c. planet

    d. lens

11. People on Earth need a telescope to see

    a. the Sun.

    b. the Moon.

    c. Neptune.

    d. Venus.

12. Which of the following is found on the Moon?

    a. phases

    b. craters

    c. atmosphere

    d. liquid water

The Good Ship Popsicle Stick

Read the Unit Literature feature in your textbook.

Write About It

Response to Literature This article is about a ship made from ice cream sticks. What words are used to describe the ship? Choose an object around you. Then use words to tell about it.

Paragraphs should have a clear topic sentence that directly

addresses the object students are writing about. The sentences that

follow the topic sentence should support the topic sentence by

providing details such as size, shape, color, smell, and feel. Students

should use a closing sentence that restates the main idea of the

paragraph. Good paragraphs will stay on topic, demonstrate vivid

word choice, and include correct grammar and mechanics.

_____

_____

_____

_____

_____

_____

Name _____ Date _____

Observing Matter

Complete the chart below to show some of the
characteristics of matter. Some answers have been
written for you.

Matter

has certain properties			
volume	color	taste	size
shape	mass	smell	

can be measured		
tool	**used to measure**	**metric unit**
thermometer	temperature	degrees Celsius
ruler	length	meter
beaker or graduated cylinder	volume	liters
pan balance	mass	grams
spring scale	weight	newtons

exists in different states		
state	**definite volume?**	**definite shape?**
solid	yes	yes
liquid	yes	No. It takes the shape of its container.
gas	No. It takes the shape of its container.	No. It takes the shape of its container.

© Macmillan/McGraw-Hill

Properties of Matter

Use your textbook to help you fill in the blanks.

What is matter?

1. Everything that takes up space is _____ matter _____ .

2. The amount of space that an object takes up is

   its _____ volume _____ .

3. A large object has more volume than a(n) _____ small _____
   object.

4. An object's _____ mass _____ is equal to the amount
   of matter it has.

5. An object that feels light, such as a beach ball, has

   a(n) _____ small _____ mass.

6. A characteristic of matter is called a(n) _____ property _____ .

What are some properties of matter?

7. Two properties of matter are volume and _____ mass _____ .

8. The size, smell, feel, and _____ shape _____ of an
   object are also properties.

9. An object will sink or float because of its _____ volume _____
   and mass.

10. An object with a small mass and a large volume will

    usually _____ float _____ .

© Macmillan/McGraw-Hill

11. An object with a large mass and a small volume will

    usually _____sink_____ .

12. A magnet pulls on objects made of _____iron_____ .

13. Metals are good material for cooking pots because

    _____heat_____ can move through them easily.

What is matter made of?

14. Matter is made of building blocks called _____elements_____ .

15. Most matter contains more than _____one_____ element.

16. The elements hydrogen and _____oxygen_____ make up water.

17. Sugar contains the elements hydrogen, oxygen,

    and _____carbon_____ .

Critical Thinking

18. What are some properties that describe the matter in a yellow pencil?

    Possible answer: A yellow pencil is usually long, made of wood,

    and painted yellow on the outside. It contains a gray material

    in the center, may have a sharp point, and makes marks when

    I write. It is usually light, so it has a small mass. It isn't big, so it

    has a small volume.

Properties of Matter

**Match each word in the box to its definition.
Write its letter in the space provided.**

a. carbon	**c.** magnet	**e.** matter	**g.** volume
b. elements	**d.** mass	**f.** property	

1. ___g___ the amount of space an object takes up

2. ___c___ something that pulls on an object made of iron

3. ___a___ the third element in sugar, along with hydrogen and oxygen

4. ___f___ the size, shape, feel, or smell of something

5. ___e___ anything that takes up space

6. ___b___ the building blocks of matter

7. ___d___ a measure of the amount of matter in an object

Properties of Matter

Fill in the blanks.

elements	mass	properties	volume
magnetism	matter	small	

Everything around you takes up space. Anything

that takes up space is made of _____matter_____ . The

amount of space an object takes up is its _____volume_____ .

The volume of an object tells how big or _____small_____

it is. For example, a bowling ball has more volume than

a tennis ball. Bowling balls are heavier than tennis balls

because bowling balls have more matter. A bowling ball

therefore has more _____mass_____ than a tennis ball.

The color, shape, feel, and smell of an object are its

_____properties_____ . Objects have many different kinds of

properties, such as _____magnetism_____ and conducting

heat. All matter is made up of building blocks called

_____elements_____ . Different combinations of elements

make up all matter.

© Macmillan/McGraw-Hill

Meet Neil deGrasse Tyson

Read the Reading in Science feature in your textbook.

Write About It

Main Idea and Details Read the article with a partner. What is the main idea? What details add to the main idea? Fill in a main-idea chart. Then write a few sentences to explain the main idea.

Use the graphic organizer to complete the main idea and supporting details found in the article.

Main Idea
Your body contains hydrogen, carbon, and many other
_____elements_____ . They formed in _____stars_____
long ago.

| Most elements form inside the _____centers of stars_____ . Hydrogen _____combines_____ to form all of the other elements. | Stars _____scatter_____ elements into space. | Over _____millions of years_____ , these elements combine to form new _____stars_____ , planets, or _____living things_____ . |

Name _____ Date _____

Planning and Organizing

Answer the questions below about the article.

1. What does Dr. Neil deGrasse Tyson study?

   how the universe works

2. Where does Dr. Tyson work?

   the American Museum of Natural History in New York

3. What is your body made up of?

   hydrogen, carbon, and many other elements

4. Where do most elements form? _____ inside the centers of stars

5. What combines to form all the other elements? _____ hydrogen

6. How do these elements make their way from the stars to your body?

   Stars explode and scatter elements into space. Over millions of

   years, these elements combine to form new stars, planets, or

   living things.

Drafting

▶ Start by writing a clear statement that describes the main idea of the article.

▶ Write three supporting details.

▶ Read what you have written. Cross out anything that does not directly support the main idea.

▶ Exchange papers with your partner and ask him or her to check your choice of a main idea. Have your partner also check your choice of supporting details.

Measuring Matter

Use your textbook to help you fill in the blanks.

How is matter measured?

1. The sizes, or amounts, of matter in objects can be compared by _____measuring_____ .

2. A unit of measurement that people agree to use is called a _____standard unit_____ .

3. Standard units of measure in the _____metric system_____ are meters, grams, and liters.

4. A thermometer is used to measure the _____temperature_____ of a substance or an object.

5. In the metric system, volume is measured in _____liters_____ .

6. Scientists use equipment such as _____beakers_____ and _____graduated cylinders_____ to measure volume.

How do we measure mass?

7. The mass of an object can be measured on a(n) _____pan balance_____ .

8. The amount of matter in an object is referred to as its _____mass_____ .

9. In the metric system, mass is measured in _____grams_____ .

10. An object with particles packed tightly together has more mass than an object in which particles

are _____far apart_____ .

How are mass and weight different?

11. The force that pulls objects to Earth is called _____gravity_____ .

12. The measure of the amount of gravity pulling an

object toward Earth is its _____weight_____ .

13. The weights of certain objects can be measured

using a(n) _____spring scale_____ .

Critical Thinking

14. Why would a brick have the same mass on the Moon as it has on Earth, but weigh less on the Moon?

The mass of the brick would not change, because the brick is

still made up of the same number of particles. The weight of

the brick, however, is affected by the pull of gravity. The pull of

gravity on Earth is more than the pull of gravity on the Moon

and therefore its weight will be greater on Earth than on

the Moon.

_____

_____

© Macmillan/McGraw-Hill

Measuring Matter

What am I?

Choose a word from the box below that answers each question.

a. gravity	**c.** meter	**e.** pan balance	**g.** spring scale
b. liter	**d.** metric system	**f.** standard unit	**h.** weight

1. ___c___ I am the unit of length in the metric system. What am I?

2. ___f___ I am a unit of measure that people agree to use. What am I?

3. ___e___ I am a tool used to measure mass. What am I?

4. ___h___ I am different on the Moon than on Earth. What am I?

5. ___g___ I am a tool used to measure weight. What am I?

6. ___d___ I am a system used by scientists to make accurate measurements of matter. What am I?

7. ___b___ I am a unit of liquid volume in the metric system. What am I?

8. ___a___ I am the force that keeps objects from floating off into space. What am I?

Name _____ Date _____

Measuring Matter

Fill in the blanks.

gravity	metric system	weight
mass	tightly	

All matter is made of small particles. Some objects contain particles that are far apart and some have many particles packed _____tightly_____ together. Therefore, an object like a bowling ball has more mass than an object like a balloon because it has more particles that are close together.

The _____mass_____ of an object on Earth is the same as it is on the Moon because the number of particles in an object stays the same. However, an object's _____weight_____ on Earth is greater than it would be on the Moon because the pull of _____gravity_____ is greater on Earth than it is on the Moon. Scientists use the _____metric system_____ to measure matter. Scientists use these measures often in their daily work.

Solids, Liquids, and Gases

Use your textbook to help you fill in the blanks.

What are three forms of matter?

1. Three forms of matter are solid, liquid, and _____gas_____ .

2. These three forms are what scientists call the _____states of matter_____ .

3. Solids, liquids, and gases each have certain _____properties_____ .

4. Matter that has a(n) ____size and shape____ is a solid.

5. Objects that are made of _____metal_____ , _____plastic_____ , and _____wood_____ are solids.

6. The particles in a solid are ____closely packed together____ .

What are liquids and gases?

7. Liquids and gases are matter because they take up space and have _____mass_____ .

8. Anything with a definite volume but not a definite shape is a(n) _____liquid_____ .

9. Milk is a liquid because it takes the _____shape_____ of its container.

10. Whether a cup of milk is spilled or in a glass, the milk still has the same _____volume_____ .

© Macmillan/McGraw-Hill

11. Particles in liquids are not as ___close together___ as particles in solids.

12. Particles in gases have more ___energy___ than particles in liquids.

13. Any matter that does not have a definite shape or volume is a(n) ___gas___ .

14. Gases spread out to take the shape and ___volume___ of their containers.

15. Particles in gases move about ___freely___ .

How do you use all the states of matter?

16. The handlebars and seat of a bicycle are ___solids___ .

17. The air in bicycle tires is a(n) ___gas___ . The chain ___oil___ is a liquid.

Critical Thinking

18. How are the three states of matter represented each day when you eat lunch?

Possible answer: Solids are represented by the food I eat, by

the table I sit at and the chair I sit on, and by the clothes I wear.

Liquids are represented by the water, milk, or juices that I drink.

Gas is represented by the air that surrounds me.

_____

© Macmillan/McGraw-Hill

Solids, Liquids, and Gases

Match the correct letter to its description.

a. definite	**d.** liquid	**g.** solid
b. volume	**e.** oxygen	**h.** states of matter
c. gas	**f.** particles	

1. ____b____ stays the same in a liquid

2. ____g____ matter that has particles packed tightly together

3. ____a____ means "it has a limit to its shape and size"

4. ____e____ a gas needed by living things

5. ____c____ matter with particles that can be far apart or squeezed together

6. ____h____ forms that scientists call gases, solids, and liquids

7. ____f____ all matter is made of these

8. ____d____ matter whose particles have no definite shape but take the shape of the matter's container

Solids, Liquids, and Gases

Fill in the blanks below using the words in the box.

definite	gas	slide past	tightly packed
energy	less	solid	
freely	liquid	spread out	

Every day, living things use substances in different

states of matter. These states of matter are _____gas_____,

_____liquid_____, and _____solid_____ .

Gases, liquids, and solids have different characteristics.

Gases have particles that are far apart from each other.

Particles in gases have a lot of _____energy_____ and

move _____freely_____ . They _____spread out_____ to fill up

whatever container they are in. Liquids are made up of

particles that have _____less_____ energy than gases.

Particles in a liquid _____slide past_____ one another. They

take the shape of their container. Solids have a(n)

_____definite_____ shape. Particles in a solid have the least

amount of energy and are _____tightly packed_____ together.

They do not move around much.

© Macmillan/McGraw-Hill

Describe Matter

Read the Writing in Science feature in your textbook.

 Write About It

Descriptive Writing Think of an object you use every day, such as your book bag. How would you describe it to someone who has never seen it before? Use the object's properties to write a description of the object.

Getting Ideas

Select one object. Write it in the center oval of the web below. Brainstorm details that describe it. Write them in the outer ovals.

Possible web:

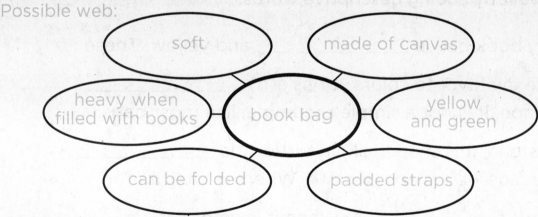

soft

made of canvas

heavy when filled with books

book bag

yellow and green

can be folded

padded straps

Planning and Organizing

Here are two sentences that Malcolm wrote about his book bag. Write "yes" if the sentence includes details that describe the bag. Write "no" if it does not.

1. My book bag is soft and crinkly. ___yes___

2. I carry my book bag to school every day. ___no___

Name _____ Date _____

Drafting

Write a sentence that begins your description. Identify the object that you are describing and the most important idea about it. This is your topic sentence.

Possible sentence: My book bag is one of the most useful things I

have.

Now write your description on a separate piece of paper. Begin with your topic sentence. Include details to help your readers picture the object.

Revising and Proofreading

Here is part of Malcolm's description. Help him improve it by adding descriptive words.

My book bag is _____green_____ and yellow. These

are my favorite colors. It has a(n) _____unusual_____

shape. It is not a simple rectangle, like most bags.

Instead, it is shaped like a turtle. It is _____soft_____

because it is made of cloth. When it is full, it is

very _____heavy_____ . When it is empty, it

is _____light_____ .

Now revise and proofread your writing. Ask yourself:

► Did I include details to describe how the object looks, sounds, feels, smells, or tastes?

► Did I put these details in an order that makes sense?

► Did I correct all mistakes?

Observing Matter

Circle the letter of the best answer.

1. What is the measure of the amount of matter in an object?

   a. weight

   b. volume

   c. **mass**

   d. gravity

2. What type of object attracts objects made of iron?

   a. gravity

   b. **magnet**

   c. gas

   d. liquid

3. In a solid, particles

   a. move freely.

   b. slide past one another.

   c. have a large amount of energy.

   d. **do not have much room to move.**

4. A standard unit in the metric system is the

   a. inch.

   b. pound.

   c. **meter.**

   d. gallon.

5. The unit used to measure liquid volume in the metric system is the

   a. meter.

   b. **liter.**

   c. centimeter.

   d. gram.

6. The amount of space an object takes up is its

   a. mass.

   b. state of matter.

   c. **volume.**

   d. temperature.

© Macmillan/McGraw-Hill

Circle the letter of the best answer.

7. Mass is the measure of

   a. the amount of matter in an object.

   b. the volume of an object.

   c. the amount of gravity pulling on an object.

   d. the weight of an object.

8. Which of the following would be different if an object were on the Moon rather than on Earth?

   a. elements

   b. weight

   c. mass

   d. volume

9. Heat moves easily through

   a. wood.

   b. metal.

   c. gases.

   d. magnetic materials.

10. Sugar is made up of the elements hydrogen, oxygen, and

    a. aluminum.

    b. carbon.

    c. iron.

    d. water.

11. The handlebars and seat of a bicycle are examples of

    a. gases.

    b. liquids.

    c. solids.

    d. elements.

12. Liquids and gases are alike because they

    a. are both solids.

    b. have no definite shape.

    c. have particles that are tightly packed together.

    d. have less energy than solids.

Changes in Matter

Complete the concept map with the information you learned about matter and the ways in which matter changes.

Changes in Matter

Physical Changes	**Chemical Changes**

A physical change is a change in the way matter

_____ looks _____ .

A chemical change is a change that creates

_____ a new type of matter _____ .

After a physical change, the matter is

_____ still the same type of _____

_____ matter _____ .

After a chemical change, the matter is

_____ different and has new _____

_____ properties _____ .

Three types of physical change are:

1. _____ change in shape _____

2. _____ change in size _____

3. _____ change in state _____

Three signs that a chemical change has taken place are:

1. _____ light and heat _____

2. _____ color change _____

3. _____ bubbles _____

Changes of State

Use your textbook to help you fill in the blanks.

What happens when matter is heated?

1. When something melts, it changes from a(n) _____solid_____

   to a(n) _____liquid_____ .

2. Matter gains _____energy_____ when it is heated.

3. Particles in solids are held _____close together_____ .

4. Particles in liquids _____flow around_____ one another.

5. When something boils, it changes from a(n) _____liquid_____

   to a(n) _____gas_____ .

6. Heat causes particles in a liquid to move _____faster_____

   and _____spread out_____ .

7. Liquids can change into a gas without boiling,

   a process known as _____evaporation_____ .

8. Water in the form of a gas is called _____water vapor_____ .

What happens when matter is cooled?

9. When a solid, a liquid, or a gas is cooled, it _____loses_____
   energy.

10. When a gas cools to the right temperature, it will

    _____condense_____ .

© Macmillan/McGraw-Hill

11. A gas that condenses loses energy and becomes

    a(n) _____ .

12. A liquid can _____ , or change into a

    solid, when it loses _____ .

How is water different from other kinds of matter?

13. Water can be a(n) _____ , a(n) _____ ,

    or a(n) _____ .

14. When water freezes, it takes up _____ space.

Critical Thinking

15. How are the particles in matter affected by getting or
    giving up energy?

    Possible answer: If heated to the right temperature, a solid's

    particles will gain enough energy so that it becomes a liquid. If

    a liquid gains enough energy, it will evaporate. Its particles will

    spread out and it will then become a gas. When gas particles

    lose energy, they will come close together and condense,

    forming a liquid. If a liquid loses enough energy, it will freeze.

    _____

    _____

    _____

Changes of State

What am I?

Choose the correct word from the box that answers each question. Write its letter in the blank provided.

a. boiling	**c.** energy	**e.** freezing	**g.** water vapor
b. condensation	**d.** evaporation	**f.** melting	

1. I am the gaseous state of water. What am I? ____g____

2. I am a process in which a liquid changes into a gas without boiling. What am I? ____d____

3. Solids, liquids, and gases have to gain or lose me in order to change phase. What am I? ____c____

4. I happen when liquids heat up and bubbles form. What am I? ____a____

5. I happen when solid matter gains energy and turns into a liquid. What am I? ____f____

6. I happen when particles of water vapor lose energy and come closer together. What am I? ____b____

7. I am the process that locks particles into position to form a solid. What am I? ____e____

© Macmillan/McGraw-Hill

Changes of State

Use the words in the box to fill in the blanks below.

condenses	expands	gas	liquids	solid
energy	freezes	heat	loses	water vapor

Most solids melt when heated to the right temperature. Once they melt, they become _____ liquids _____ . They melt because they gain _____ energy _____ in the form of _____ heat _____ . With enough heat, particles in liquids will move faster and spread apart. When a liquid boils, it evaporates and changes to a(n) _____ gas _____ .

Water exists in three states: solid, liquid, and gas. Water that has changed to a gas is called _____ water vapor _____ . When water vapor cools, it _____ loses _____ energy. It _____ condenses _____ and becomes liquid water again. When liquid water loses enough energy, it _____ freezes _____ . Particles that are frozen are locked in position and form a(n) _____ solid _____ . When water freezes, it _____ expands _____ . Empty spaces form between the particles and cause the water to take up more space.

Physical Changes

Use your textbook to help you fill in the blanks.

What are physical changes?

1. A change in the appearance of matter is a(n) __physical change__ .

2. Matter may __look different__ after a physical change. However, it is still the __same kind__ of matter.

3. Examples of physical changes are tearing, stretching, __cutting__ , and __bending__ .

4. Three kinds of physical changes are changes in __size__ , changes in __shape__ , and changes in __state__ .

What happens when you mix matter?

5. When different kinds of matter are put together, a(n) __mixture__ forms.

6. In a mixture, the __properties__ of each part of the mixture do not change.

7. When matter is mixed evenly with another kind of matter, a(n) __solution__ forms.

8. A solution is one kind of __mixture__ .

9. An example of a solution is __salt water__ .

© Macmillan/McGraw-Hill

10. Even if you stir them for a long time, sand and water

will never form a(n) _____ mixture _____ .

11. Not all solutions contain _____ liquids _____ .

12. Air is a solution of different _____ gases _____ .

13. Several different solids make up _____ brass _____ .

How can mixtures be separated?

14. Mixtures can be separated using _____ properties _____
such as size, shape, and color.

15. In a salt-water mixture, salt is separated when

water _____ evaporates _____ .

Critical Thinking

16. Describe two ways in which matter can be changed
physically. Use paper and water as examples.

Matter can change physically, but it will always have the same
_____

composition. For example, when a sheet of paper is torn or bent,
_____

it may have a different shape or size, but it is still paper. When
_____

liquid water freezes or an ice cube melts, the state of the water
_____

is changed, but it is still water.
_____

_____

Physical Changes

Match the correct word with its description. Write in the correct letter in the blank provided.

a. evaporation	**d.** properties	**g.** solution
b. mixture	**e.** shape	**h.** state
c. physical change	**f.** size	

1. ___e___ square, round, or irregular

2. ___f___ large or small

3. ___c___ a change in the way matter looks

4. ___h___ solid, liquid, or gas

5. ___g___ one or more kinds of matter mixed evenly in another kind of matter

6. ___d___ the volume, mass, look, smell, feel, and sound of something

7. ___a___ what happens when a gas forms slowly from a liquid

8. ___b___ different kinds of matter mixed together

Physical Changes

Use the words in the box to fill in the blanks below.

gases	mixture	shape	solids
matter	properties	size	solution

Matter can undergo physical changes. Afterwards, it

may look different, but it is still the same _____matter_____ .

Tearing a sheet of paper changes its _____shape_____ ,

but it is still paper. Any change in the _____size_____ ,

shape, or state of matter is a physical change.

When different kinds of matter are mixed together,

a(n) _____mixture_____ results. Each kind of matter in a

mixture keeps its _____properties_____ . A mixture in

which matter mixes evenly with another kind of matter is

called a(n) _____solution_____ . Some solutions, like air,

are mixtures of _____gases_____ . Other solutions, such

as brass, are mixtures of _____solids_____ . Even two

foods, such as spaghetti and meatballs, can be called

a mixture!

Name _____ Date _____

Mining Ores

Read the Reading in Science feature in your textbook.

Write About It

Infer Read the article with a partner. Use what you know and what you read in the article to answer this question. Why do you think it is important for people to recycle metals? Write a paragraph to share your ideas.

Use the graphic organizer below to identify what you already know and what you can infer from the passage about obtaining useful metals.

Clues	What I Know	What I Infer
I use _____ metals _____ every day.	Spoons and bikes _____ are made of metals.	I can recognize metals because they are _____ used in _____ familiar objects ___.
Metals come from _____ Earth ____.	Metals are found in _____ ores ____.	I infer that there are many ways to separate metals from _____ ores ____.

Planning and Organizing

Answer the questions to help you write your essay.

1. What does recycling mean?

Recycling means taking a product that would otherwise be

thrown away and breaking it down into parts that can be used

to make new products.

2. What do you think happens to mountains or the ground where ores are mined?

They are flattened because all of the soil has been removed.

3. According to the information in the article, what things might be saved if metals are recycled?

Mountains, river valleys, and volcanoes will be preserved.

Drafting

Write two or three reasons for recycling metals. Have your partner read your work. Does your partner agree or disagree with your reasons? Why?

1. _____

_____

_____

2. _____

_____

_____

Chemical Changes

Use your textbook to help you fill in the blanks.

What are chemical changes?

1. Rust and ash are results of __chemical changes__ .

2. A chemical change takes place when a material forms

   a(n) _____new kind_____ of matter.

3. The ____properties____ of a new material will be

   different from the original material.

4. Chemical changes break down _____food_____

   in our bodies.

5. Our bodies get _____energy_____ when food is

   broken down.

6. Plants and _____animals_____ stay alive because of

   chemical changes.

7. Energy from the Sun is used by ____green plants____ .

8. Plants change carbon dioxide and water into food

   and _____oxygen_____ .

9. Rust on a car shows that chemical changes have

   happened to parts made of _____iron_____ .

© Macmillan/McGraw-Hill

What are the signs of a chemical change?

10. There are often _____signs_____ that a chemical change has taken place.

11. Signs of chemical change are _____light_____ , _____heat_____ , or _____color_____ change.

12. When a log burns, ___heat and light___ are released and ___new kinds of matter___ form.

13. Some chemical changes also make _____bubbles_____ .

14. When baking soda and vinegar are mixed, bubbles of ___carbon dioxide___ gas form.

Critical Thinking

15. Give an example of a chemical change you have seen that is helpful, and one that is damaging.

Possible answer: A campfire is an example of a positive

chemical change. Cucumbers that have become rotten after

being left in the refrigerator for too long is an example of a

damaging chemical change.

_____

_____

_____

Chemical Changes

Match the correct word with its description. Then write its letter in the blanks below.

a. baking	**d.** chemical change	**g.** properties
b. bubbles	**e.** green plants	**h.** rust
c. carbon dioxide	**f.** light and heat	

1. ____f____ These are two signs that a chemical change has taken place.

2. ____c____ Plants change this material into food and oxygen.

3. ____d____ This process goes on inside you every day and creates a new type of matter.

4. ____g____ These are the characteristics of a certain type of matter.

5. ____a____ During this activity, cake batter is changed chemically.

6. ____h____ When iron is chemically changed, this is made.

7. ____e____ These use the Sun's energy to make chemical changes.

8. ____b____ These show that carbon dioxide gas forms when baking soda is added to vinegar.

© Macmillan/McGraw-Hill

Chemical Changes

Use the words in the box to fill in the blanks below.

breaks down	color	light	properties
bubbles	heat	new type	signs

Every day, chemical changes take place in and

around you. For example, food that your body

___breaks down___ undergoes a chemical change. When

something changes chemically, it becomes a ___new type___

of material and has different ___properties___ from the

original material.

The process of a chemical change can be detected

by certain ___signs___ , or evidence. For

instance, ___light___ and ___heat___ are

given off as logs burn. Another sign of a chemical

change is a change in ___color___ , such as when

an apple turns brown. Chemical changes may also give

off ___bubbles___ when some materials are mixed

together. The rust on a bicycle and cooked food are

also examples of a chemical change.

© Macmillan/McGraw-Hill

Changes in Matter

Circle the letter of the best answer.

1. Peas and carrots together in a bowl are an example of a(n)

   a. chemical change.

   b. solution.

   c. mixture.

   d. property.

2. Water vapor collects on the outside of a cold glass. This is an example of

   a. boiling.

   b. condensation.

   c. freezing.

   d. melting.

3. When a liquid evaporates, it

   a. becomes a solid.

   b. changes color.

   c. boils.

   d. changes state.

4. When a string is cut into two pieces, a(n) _____ has occurred.

   a. chemical change

   b. change in state

   c. physical change

   d. change in the string's properties

5. What can happen when two materials are combined to make a new type of matter?

   a. The shapes of the materials change.

   b. The sizes of the materials change.

   c. Heat and light are released.

   d. The materials can be separated easily.

© Macmillan/McGraw-Hill

Circle the letter of the best answer.

6. What new types of matter do green plants produce with the help of the Sun's energy?

   a. carbon dioxide and water vapor

   b. rust and iron

   c. food and oxygen

   d. bubbles and a color change

7. What forms when one or more types of matter are mixed evenly with another type of matter?

   a. bubbles

   b. a solution

   c. heat and light

   d. a color change

8. What happens when solid water begins to gain energy?

   a. it freezes

   b. it boils

   c. it evaporates

   d. it melts

9. A person put a plastic bottle of water in the freezer. Later, the person observed that the bottle had cracked. What property of water made this happen?

   a. The water gained energy as it froze.

   b. The water expanded, or took up more space, as it froze.

   c. The water condensed on the outside of the bottle as it froze.

   d. The water gave off bubbles of carbon dioxide as it froze.

10. One sign of a chemical change in a banana is a change in

   a. state.

   b. shape.

   c. size.

   d. color.

Jump Rope
by Rebecca Kai Dotlich

Read the Unit Literature feature in your textbook.

Write About It

Response to Literature This poet uses rhythm and rhyme to describe how a jump rope moves. How else do things move on the playground? Write a poem about another movement game.

Students' poems should demonstrate vivid word choice and include

correct grammar and mechanics. Their poems should include their

own observations of how things move in another movement game.

_____

_____

_____

_____

_____

_____

_____

_____

Forces and Motion

Complete the concept map with the information that you learned about forces and motion.

Motion

Forces

An object's
_____position_____ is the
place where the object is.

A(n) _____force_____
is a push or a(n)
_____pull_____ .

Object

When an object's
position is changing,
the object is in
_____motion_____ .

Forces can
change an object's
_____motion_____
by changing its
_____speed_____ or
direction.

The _____speed_____ of
an object describes how
fast the object is moving.

A(n) _____machine_____
can help people do
work. Some machines
let people use less
_____force_____ to do
a job.

Position and Motion

Use your textbook to help you fill in the blanks.

How can you describe position?

1. The location of an object is its _____ position _____ .

2. The words _____ over _____ , *under*, _____ left _____ ,

   right, _____ on top of _____ , and *next to* help describe the
   position of an object.

3. Position words tell where an object is by _____ comparing _____
   it to the locations of other objects.

4. Measuring is a way to find the _____ distance _____
   between objects.

5. Distance can be measured with a(n) _____ ruler _____ .

6. Distance can be used to help describe the _____ position _____
   of an object.

What is motion?

7. When an object's _____ position _____ is changing, it is
   in motion.

8. An object that is in motion can move _____ quickly _____

   or _____ slowly _____ .

9. A(n) _____ swinging _____ object moves back and forth.

10. Short, sharp turns from one side to another form

    a(n) _____ zigzag _____ path.

© Macmillan/McGraw-Hill

What is speed?

11. How fast an object is moving is described by

its _____*speed*_____ .

12. The speed of a moving object can be _____*measured*_____ .

13. To measure an object's _____*speed*_____ ,
you must know how far the object traveled and how
long it took to go that distance.

14. If you ride a bicycle 15 kilometers in one hour, you are

moving at a speed of _____*15*_____ kilometers
per hour.

Critical Thinking

15. Choose a group of objects in your classroom or in
one room of your house and use position words to
describe their relationship to each other. Use at least
three objects and three different position words.

Possible answer: In my classroom, the clock is hanging over the

blackboard. To the left of the blackboard is a bulletin board.

There is a waste basket beneath the bulletin board.

_____

_____

_____

_____

© Macmillan/McGraw-Hill

Position and Motion

Matching

Match the words in the box to their descriptions below. Write the correct letter in the space provided.

a. distance	**d.** position word	**g.** straight line
b. motion	**e.** ruler	**h.** zigzag
c. position	**f.** speed	

1. ____c____ the location of an object

2. ____h____ a path with short, sharp turns from side to side

3. ____f____ a description of how fast an object moves

4. ____e____ an item used to measure distance

5. ____d____ a term such as *right, next to,* or *under*

6. ____a____ the amount of space between two places or objects

7. ____b____ a change in position

8. ____g____ a path with no turns

© Macmillan/McGraw-Hill

Position and Motion

Use the words in the box to fill in the blanks below.

comparing	motion	straight line	zigzag
distance	slowly	under	

If you describe the position of an object, you are describing where it is. An object's position can be described by ____comparing____ it with the things near it. Words such as *over*, ____under____ , and *on top of* are useful for describing position. You might also describe an object's ____distance____ based on the things around it.

An object that is changing position is in ____motion____ . Objects, such as a bicycle, can move quickly or ____slowly____ . A bicycle can move in a(n) ____straight line____ , in a circular pattern, or in a(n) ____zigzag____ pattern. The speed of an object tells how fast the object is moving.

Travel Through Time

Read the following passage. Underline the sentences that describe new inventions. Circle the sentences that describe the achievements of those inventions.

1804 In England Richard Trevithick built the first steam engine for a train. The steam engine helped people travel great distances. It also helped them get to their destinations more quickly.

1884 In Germany Karl Freidrich Benz built the first car to run on gasoline. It worked similarly to the cars you see on the road today. However, his car only had three wheels!

1903 Wilbur and Orville Wright constructed the first motorized airplane that flew and landed safely. Their airplane's engine ran on gasoline. It flew for 12 seconds over 36 meters (120 feet).

1961 Russian astronaut Yuri Gagarin was the first person in space. His spaceship had special engines. They produced a force that was stronger than the pull of Earth's gravity. These engines helped the spaceship leave Earth's surface and orbit the planet.

Problem and Solution

Fill in the problem-and-solution graphic organizers.
Use the sentences you underlined and circled as clues.

Name _____ Date _____

Problem

Before the 1800s it took people a long time to travel great distances.

↓

Steps to Solution

In 1804, ___Richard Trevithick___ built

the first ___steam engine___ for

a(n) ___train___ .

↓

Solution

The ___steam engine___ helped

people travel ___long___

distances and reach their

destinations ___quickly___ .

Problem

Before 1961, people could not travel in space.

↓

Steps to Solution

Russian scientists built a(n)

___spaceship___ with special

engines that were stronger than

the ___force of gravity___ .

↓

Solution

Russian astronaut

___Yuri Gagarin___ was the first

person

in ___space___ .

Write About It

Problem and Solution How have machines helped people learn about distant places? Read the article again. On a separate piece of paper, write about ways machines have helped people solve problems.

Forces

Use your textbook to help you fill in the blanks.

What are forces?

1. To make an object start moving, a(n) _____force_____ must be applied to it.

2. A force that makes something move can be a(n) _____push_____ or a(n) _____pull_____ .

3. More force is needed to move _____heavy_____ objects than to move light objects.

4. Forces can make objects start moving, _____speed up_____ , _____slow down_____ , or stop moving.

5. Forces can change the _____direction_____ of a moving object.

6. When the forces of an object cancel out, like a rope being pulled equally from each side, the forces are _____balanced_____ .

What are types of forces?

7. Forces that happen between objects that touch are _____contact forces_____ .

8. Forces such as _____magnetism_____ and _____gravity_____ can act on an object without touching it.

© Macmillan/McGraw-Hill

9. Magnets can _____ attract or _____ repel
one another without touching.

10. Magnets attract or repel through _____ solids ,
_____ liquids , or _____ gases .

11. The pulling force between two objects is
called _____ gravity .

12. A measure of the pull of gravity on an object is
its _____ weight .

What is friction?

13. The force that occurs when one object rubs against
another object is called _____ friction .

14. There is very little friction between _____ smooth ,
slippery surfaces, and a lot of friction between
_____ rough surfaces.

Critical Thinking

15. In a baseball game, the first batter hit the ball far
into the outfield. The second batter did not hit the
ball as far and the ball only made it onto the infield.
How do you know that the first batter used more
force on the ball?

Possible answer: The first batter used more force because the

ball went much farther. It takes more force to hit a ball into the

outfield than it does to bunt a ball into the infield.

Name _____ Date _____

Forces

**Match each word in the box with its definition below.
Write the correct letter in the space provided.**

a. balanced	**c.** friction	**e.** magnet	**g.** unbalanced
b. force	**d.** gravity	**f.** repel	**h.** weight

1. ___a___ forces on an object that cancel each other out because they have equal and opposite effects

2. ___b___ a push or a pull

3. ___f___ to push away

4. ___c___ a force that occurs when objects rub against each other

5. ___h___ a measure of the amount of gravity between two objects

6. ___e___ an object with magnetic force

7. ___g___ when forces on an object do not cancel each other out

8. ___d___ a pulling force between two objects, such as between you and Earth

Forces

Use the words in the box to fill in the blanks below.

contact forces	friction	rough
direction	gravity	slippery
force	push	touching

What makes a soccer ball move? To make any

object move, a(n) _____force_____ has to be applied

to it. The force may be a(n) _____push_____ or a

pull. In soccer, the goalie's job is to use force to stop

or change the _____direction_____ of the ball. The goalie

uses _____contact forces_____ to do this.

The force that works against motion when one

object rubs another object is _____friction_____ .

Surfaces that are _____rough_____ have more friction

than surfaces that are _____slippery_____ . Magnets

attract or repel each other without _____touching_____ .

The force of _____gravity_____ also can pull objects

from a distance. For example, gravity is the force that

pulls you toward Earth.

Work and Energy

Use your textbook to help you fill in the blanks.

What is work?

1. In science, _____*work*_____ is done when a force
   changes the motion of an object.

2. Picking up a book is work because a force is used
   and the book _____*moves*_____ .

3. Dropping a book is work because _____*gravity*_____
   changes the motion of the book.

4. Work is not done on an object when you push it and
   it does not _____*move*_____ .

5. Picking up a boulder is _____*more*_____ work than
   picking up a pebble.

What is energy?

6. People need _____*energy*_____ to do work.

7. The ability to do _____*work*_____ is a way to
   describe _____*energy*_____ .

8. Energy makes it possible for changes in _____*motion*_____
   to take place.

9. The two main forms of energy are _____*potential energy*_____
   and _____*kinetic energy*_____ .

© Macmillan/McGraw-Hill

10. A flying plane has kinetic energy because it is _____ *moving* .

11. Energy that is stored is _____ *potential energy* .

12. A ball at the top of a hill has _____ *potential energy* because of its position.

13. As the ball rolls down the hill, its potential energy changes into _____ *kinetic energy* .

14. When wood or food is burned, its _____ *potential energy* changes into kinetic energy.

How can energy change?

15. Energy can change from one _____ *form* to another.

16. You _____ *transfer* energy from your body to a bowling ball when you roll the ball down an alley.

Critical Thinking

17. Use the terms *potential energy* and *kinetic energy* to describe what happens when a car uses gasoline.

Possible answer: Gasoline has potential energy stored in it.

As the car moves, its gasoline burns and its stored energy is

changed into kinetic energy.

_____

_____

Work and Energy

Matching

Match each word in the box with its definition below.
Write the letter of the word in the space provided.

a. energy	**d.** heat	**g.** work
b. friction	**e.** kinetic energy	
c. gravity	**f.** potential energy	

1. ____d____ you feel this when you rub your hands together

2. ____f____ energy that is ready to be used

3. ____b____ a force that works to slow your hands when you rub them together

4. ____e____ energy in running water

5. ____a____ the ability to do work

6. ____g____ when a force changes an object's motion

7. ____c____ a force that causes an object on Earth to fall

Work and Energy

Use the words in the box to fill in the blanks below.

drop	gravity	motion	stored energy
energy	kinetic energy	moved	work

In science, work is done only when a force changes the motion of an object. If you pick up a pencil from the floor, you have done _____**work**_____ . That is because a force _____**moved**_____ the pencil. If you _____**drop**_____ the pencil, more work has been done. By dropping the pencil, you allowed the force of _____**gravity**_____ to change the pencil's motion.

In order to do work, you need _____**energy**_____ . Energy is the ability to do work. Energy is what makes it possible to change the _____**motion**_____ of objects. The two main forms of energy are potential energy and _____**kinetic energy**_____ . Potential energy is _____**stored energy**_____ that is ready to be used. Anything that moves, such as running water or a rolling ball, has kinetic energy.

Name _____ Date _____

Using Simple Machines

Use your textbook to help you fill in the blanks.

What are machines?

1. An object that makes work easier is a(n) _____machine_____ .

2. Machines with few or no moving parts are called
   _____simple_____ machines.

3. Six types of simple machines are the lever, the
   _____pulley_____ , the wheel and axle, the _____inclined plane_____ ,
   the wedge, and the _____screw_____ .

What are levers?

4. A straight bar that moves on a fixed point, such as a
   seesaw, is a(n) _____lever_____ .

5. An object that is lifted by a machine is the _____load_____ .

6. A(n) _____pulley_____ uses a rope and wheel to lift
   a load.

7. A wheel that moves around a post is a lever called
   the _____wheel and axle_____ .

What are inclined planes?

8. A simple machine with a flat, _____slanted_____
   surface is an inclined plane.

9. Inclined planes _____reduce_____ the force needed to move an object, but the object has to be pushed a(n) _____longer_____ distance.

10. A(n) _____wedge_____ is a simple machine that uses force to split an object apart.

11. A(n) _____screw_____ is an inclined plane that is wrapped into a spiral.

How do machines work together?

12. Two or more simple machines working together make a(n) _____compound machine_____ .

13. A can opener is a compound machine made up of a(n) _____wedge_____ , a(n) _____lever_____ , and a wheel and axle.

Critical Thinking

14. What simple machine could you use to lift a heavy rock out of the ground? How would you use the machine?

    Possible answer: I could use a crowbar as a lever. I could put one end of the crowbar under the rock, place another rock beneath the bar as a fulcrum, and then push down on the opposite end of the bar to lift the rock.

Name _____ Date _____

Using Simple Machines

What am I?

Choose a word from the box that answers each
question below. Write its letter in the space provided.

a. compound machine	**d.** lever	**g.** screw
b. fulcrum	**e.** load	**h.** simple machine
c. inclined plane	**f.** pulley	

1. ____f____ I am a kind of lever that uses a rope and wheel to lift an object. What am I?

2. ____b____ I am a fixed point on a lever. What am I?

3. ____c____ I am a simple machine with a flat, slanted surface. What am I?

4. ____a____ I am made up of two or more simple machines. What am I?

5. ____d____ I am a straight bar that moves on a fixed point. What am I?

6. ____e____ I am the object lifted by a lever. What am I?

7. ____g____ I am an inclined plane wrapped into a spiral. What am I?

8. ____h____ I am a lever, pulley, wheel and axle, inclined plane, wedge, or screw. What am I?

Using Simple Machines

Use the words in the box to fill in the blanks below.

compound machines	inclined plane	pulley
force	lever	six
fulcrum	moving	wheel and axle

Every day, you use machines that make work easier

to do. A machine helps you use less _____force_____

to do a job. A machine can also change the direction of

the force you use. Most tools are ___compound machines___

made up of two or more simple machines.

There are _____six_____ types of simple

machines. Simple machines have few or no ____moving____

parts. A seesaw is an example of a(n) _____lever_____ .

It is a straight bar that moves on a fixed point, called

a(n) ____fulcrum____ , to lift a load. A(n) ____pulley____

uses a rope and wheel to lift a load. A doorknob is an

example of a(n) ___wheel and axle___ . When you slide a

box up a ramp, you are using a(n) ___inclined plane___ .

A knife is a wedge that splits food apart.

Name _____ Date _____

A Very Useful Machine

Read the Writing in Science feature in your textbook.

 Write About It

Explanatory Writing Choose another compound machine. Find out how it works. Then write a paragraph that explains how to use it.

Getting Ideas

Think about how to use the machine you chose and how it works. Then write the steps in the chart below.

Name of Machine: _____zipper_____

First
Pull the slide up the zipper.

↓

Next
The hooks in the teeth hold the two sides together.

↓

Last
The wedges in the slide force the teeth together.

Planning and Organizing

Eric wrote about how a zipper works. Here are two sentences that he wrote. Write "yes" if the sentence tells how the machine works. Write "no" if it does not.

1. Its slide is made up of wedges that are inclined. ____yes____

2. The wedge pushes the edge of the hooks to open them. ____no____

© Macmillan/McGraw-Hill

Drafting

Write a topic sentence. State what compound machine you are writing about. Write an important idea about it.

Possible sentence: The zipper uses wedges and hooks to fasten things.

Now write your explanation. Use a separate piece of paper. Start with your topic sentence. Then describe how the machine works. Write the steps in order. Use time order words to make the steps easy to follow.

Revising and Proofreading

Here are some sentences that Eric wrote. They explain how a zipper works. Combine each pair of sentences using the time order word in parentheses. Write the new sentence on the line provided.

1. The teeth connect. You pull the slide up the zipper. (as)

   The teeth connect as you pull the slide up the zipper.

2. The wedge forces the teeth apart. You pull the slide down. (when)

   The wedge forces the teeth apart when you pull the slide down.

Now revise and proofread your writing.
Ask yourself:

▶ Did I explain how the machine works?

▶ Did I write the steps in order?

▶ Did I correct all mistakes?

Forces and Motion

Circle the letter of the best answer.

1. The fixed point at which a lever moves is the

   a. load.

   (b.) fulcrum.

   c. axle.

   d. pulley.

2. A bus traveled 35 kilometers in one hour. What was the speed of the bus?

   a. 70 kilometers per hour

   b. 10 kilometers per hour

   c. 25 kilometers per hour

   (d.) 35 kilometers per hour

3. Potential energy is changed into kinetic energy when

   a. a bicycle slows down.

   b. a book rests on a table.

   (c.) a sled moves down a hill.

   d. paper blows in the wind.

4. When one team in a tug of war pulls harder on the rope than the other team does, the forces are

   a. active.

   b. balanced.

   c. magnetic.

   (d.) unbalanced.

5. The position of an object is its

   a. age.

   b. color.

   (c.) location.

   d. size.

6. The force that occurs when one object rubs against another object is

   (a.) friction.

   b. gravity.

   c. magnetism.

   d. weight.

Circle the letter of the best answer.

7. The ramp on the back of a truck is a simple machine called a(n)

  a. screw.

  b. pulley.

  c. inclined plane.

  d. wheel and axle.

8. When you rub your hands together, some energy of motion is changed to

  a. energy of position.

  b. heat.

  c. stored energy.

  d. water.

9. If you want to measure the distance between two books on your desk, you would use a(n)

  a. barometer.

  b. ruler.

  c. clock.

  d. thermometer.

10. An example of a change in motion caused by a contact force is

  a. gravity pulling you toward Earth.

  b. two magnets pushing each other apart.

  c. hitting a baseball with a bat.

  d. a book falling from your desk.

11. A simple machine that causes sideways forces and splits an object apart is a(n)

  a. inclined plane.

  b. wedge.

  c. lever.

  d. wheel and axle.

12. Work is done when a force changes an object's

  a. mass.

  b. color.

  c. motion.

  d. weight.

Forms of Energy

Complete the concept map with information you
have learned. Some of the answers have been
written for you.

Heat

1. Heat is the flow of
   _____thermal energy_____ from a(n)
   _____warmer_____ object
   to a cooler object.

2. The _____Sun_____ is
   Earth's main source of
   heat energy.

Sound

1. Sound begins when
   something
   _____vibrates_____ , or
   moves back and forth
   quickly.

2. Volume is how
   _____loud_____ a
   sound is.

Forms of Energy

Light

1. Light is energy that allows
   you to _____see_____ .

2. The light that we see
   is _____reflected_____
   from objects to our
   _____eyes_____ .

Electricity

1. Electricity is made up of
   particles that are neither
   _____positive_____ nor
   _____negative_____ .

2. A(n) _____circuit_____ is
   a complete path through
   which electricity can flow.

Heat

Use your textbook to help you fill in the blanks.

What is heat?

1. Heat always flows from a(n) _____warmer_____ object

   to a(n) _____colder_____ one.

2. Earth's main source of heat is the _____Sun_____ .

3. Heat can move through _____solids_____ ,

   _____liquids_____ , _____gases_____ , and space.

How does heat affect matter?

4. Particles in a(n) _____cold_____ object have little

   thermal energy and move _____slowly_____ .

5. Particles in a(n) _____hot_____ object have a lot
   of thermal energy and move quickly.

6. A measure of thermal energy, or how hot or cold

   something is, is called _____temperature_____ .

7. When an object _____gains_____ thermal energy, it

   expands and becomes _____larger_____ .

8. When an object loses thermal energy, it _____contracts_____

   and becomes _____smaller_____ .

9. The instrument used to measure temperature is a

   liquid-filled tube called a(n) _____thermometer_____ .

10. The liquid in a thermometer _____expands_____ and
    rises when the material around it gets warmer,
    and the liquid contracts and falls when the

    temperature _____decreases_____ .

How can you control the flow of heat?

11. Any material through which heat moves easily, such

    as a metal pot, is a(n) _____conductor_____ .

12. Any material through which heat does not move

    easily is a(n) _____insulator_____ .

Critical Thinking

13. How do you know that Earth is cooler than the Sun?

    Possible answer: Thermal energy or heat always moves from a

    warmer object to a cooler object. Heat moves from the Sun to

    Earth, so Earth must be cooler than the Sun.

    _____

    _____

    _____

Heat

Match the words in the box to their definitions below.

a. conductor	**d.** insulator	**g.** thermometer
b. contract	**e.** Sun	
c. expand	**f.** thermal energy	

1. ___f___ energy that makes particles in materials move

2. ___c___ to get bigger

3. ___b___ to get smaller

4. ___e___ Earth's main source of heat

5. ___d___ a material through which heat does not move easily

6. ___g___ a tool used to measure temperature

7. ___a___ a material through which heat moves easily

Name _____ Date _____

Heat

Use the words in the box to fill in the blanks below.

conductor	quickly	thermal energy
loses	smaller	thermometer

Temperature is a measure of how hot or cold

something is. It can be measured with a(n) ___thermometer___ .

Temperature tells how much ___thermal energy___ an

object has. A high temperature means that the particles

in an object have a lot of thermal energy and are

moving ___quickly___ . A low temperature means

that the particles in an object are moving slowly.

When particles in an object move faster, the object

expands. When an object ___loses___ energy, it

gets ___smaller___ , or contracts. Any material

through which heat moves easily is a(n) ___conductor___ .

Heat always travels from a hotter object to a

colder object.

© Macmillan/McGraw-Hill

Sound

Use your textbook to help you fill in the blanks.

What is sound?

1. Sound is produced when an object moves back and forth quickly, or _____vibrates_____ .

2. Sound happens only when something _____moves_____ .

3. When a sound is made, vibrations move through the air in _____waves_____ in all directions.

4. Sound travels through all types of matter, but at different _____speeds_____ .

5. Sound travels slowest through a(n) _____gas_____ . Sound travels more quickly through _____liquids_____ and most quickly through _____solids_____ .

How are sounds different?

6. A _____pitch_____ is how high or low a sound is.

7. The speed of a(n) _____vibration_____ tells whether a sound will be a high pitch or a low pitch.

8. The _____length_____ of a musical instrument's strings affects pitch. An object's _____thickness_____ also affects the speed at which it vibrates.

9. The loudness of a sound is its _____volume_____ .

10. An object that vibrates with a lot of _____ energy _____ is loud.

How do you hear sounds?

11. Vibrations in the air are collected by your _____ outer ear _____ .

The vibrations make your _____ eardrum _____ move back and forth.

12. Your vibrating eardrum makes three _____ tiny bones _____ in your ear begin to vibrate.

13. The bones pass the vibrations to the _____ inner ear _____ ,

where _____ nerves _____ send a message to your brain.

14. Loud sounds cause _____ hearing loss _____ because they

carry so much _____ energy _____ .

Critical Thinking

15. What would happen if the bones in your ears could not vibrate?

If the bones in my ears could not vibrate, they wouldn't pass the

vibrations to the nerves in my inner ear. If the nerves in my ear

could not pass information to my brain, I would not be able to

hear sounds.

_____

_____

© Macmillan/McGraw-Hill

Sound

What am I?

Choose a word from the box below that answers each question. Write its letter in the space provided.

a. eardrum	**c.** outer ear	**e.** three bones	**g.** volume
b. inner ear	**d.** pitch	**f.** vibration	**h.** wave

1. ____d____ I am how high or low a sound is. What am I?

2. ____e____ I pass vibrations to nerves in the inner ear. What am I?

3. ____g____ I am how loud a sound is. What am I?

4. ____c____ I collect sounds. What am I?

5. ____h____ I am the way that sound travels out in all directions. What am I?

6. ____a____ I make three tiny bones vibrate. What am I?

7. ____f____ I am a quick back-and-forth motion. What am I?

8. ____b____ I am the place where vibrations make nerves send messages to the brain. What am I?

Sound

Use the words in the box to fill in the blanks below.

eardrum	inner ear	outer ear	speed
high-energy	nerves	pitch	waves

Sound is produced when an object vibrates, or

moves back and forth quickly. Sound ___waves___

move out in all directions and reach your ear. Your

___outer ear___ collects these vibrations. They

make your ___eardrum___ vibrate and move three

tiny bones inside your ear. These movements cause

___nerves___ in the ___inner ear___ to send

messages to your brain, and you hear sound.

A sound's ___pitch___ may be high or low.

Pitch depends on the ___speed___ of the

vibration. A ___high-energy___ vibration will cause

a louder sound than a low-energy vibration. Sound

travels at many speeds and through a variety of

materials.

© Macmillan/McGraw-Hill

Light

Use your textbook to help you fill in the blanks.

What is light?

1. The form of energy that allows you to see objects

   is _____*light*_____ .

2. Some sources of light are _____*the Sun*_____ , fire,

   and _____*light bulbs*_____ .

3. Light travels in a(n) _____*straight path*_____ .

4. Light strikes an object and bounces, or _____*reflects*_____ ,
   off of it. You see an object when light reflects from

   it and hits your _____*eyes*_____ .

What happens when light hits different objects?

5. Materials that block the passage of light are _____*opaque*_____ .

6. The dark space formed by an object that blocks light

   is a(n) _____*shadow*_____ .

7. Materials through which light passes are _____*transparent*_____ .

8. Materials are _____*translucent*_____ when they block
   some of the light but allow the rest of it to pass
   through.

Name _____ Date _____

Why can you see colors?

9. Light from the Sun is a(n) _____mixture_____ of different colors.

10. A piece of glass called a(n) _____prism_____ separates white light into the colors that make it up.

11. An object looks black when _____all_____ light is absorbed. It looks white when all light is _____reflected_____.

How do you see?

12. You can see because light passes through your _____cornea_____ into an opening called the _____pupil_____.

13. Light is then refracted by the _____lens_____ onto the back of your eyeball.

14. Then the _____optic nerve_____ carries information to the brain and you see a picture of an object.

Critical Thinking

15. What facts about light allow you to see that a ball is blue with white stripes?

Parts of the ball are absorbing all colors but blue, so you see

reflected blue stripes. Some areas absorb no energy. These

parts reflect all the energy, so you see them as white.

© Macmillan/McGraw-Hill

Light

Match the correct word with its definition. Write its letter in the space provided.

a. absorbed	**c.** opaque	**e.** reflected	**g.** shadow
b. light	**d.** prism	**f.** refract	**h.** translucent

1. ___f___ to pass light from one material to another

2. ___b___ the form of energy that allows you to see objects

3. ___c___ the types of materials that block light from passing through them

4. ___h___ the types of materials that block some light and allow some light through them

5. ___a___ when some or all light is taken in

6. ___d___ a piece of glass that refracts light

7. ___e___ when light bounces off an object

8. ___g___ a dark space formed by opaque materials that block light from passing through them

Light

Use the words in the box below to fill in the blanks.

energy	light	reflected	translucent
eyes	opaque	refracted	transparent

There are many different types of energy. The

form of _____energy_____ that lets you see objects is

_____light_____ . Light hits an object and is _____reflected_____

from the object to your eyes. Light can be bent, or

_____refracted_____ , as it goes from one material to

another.

Light does not pass through every material.

Materials that block light are _____opaque_____ .

Materials that allow light to pass through them are

_____transparent_____ . Those that let only some light

energy through them are _____translucent_____ . You

see objects as light is reflected from them to your

_____eyes_____ . The brain uses this information to

create a picture.

© Macmillan/McGraw-Hill

A Beam of Light

Read the paragraph below.

Surgeons are doctors who perform operations to fix injuries or treat diseases. They can use scalpels—special tools with sharp blades—to cut through skin, muscles, and organs of the human body. Today, surgeons have another tool they can use to do operations. This tool is a beam of light!

This beam of light is called a *laser*. Lasers are very powerful. They can cut through the human body without causing much bleeding.

Lasers were first used to remove birthmarks on children's skin. Today, surgeons also use lasers to treat injuries to the brain, the heart, and many other parts of the body. Lasers are also used to improve people's eyesight.

Write About It

Summarize Read the article again. List the most important information in a chart. Then use the chart to summarize the article.

Planning and Organizing

▶ List the most important information from the article in the chart below.

Most Important Information

Drafting

▶ Start by writing a clear statement that describes the main idea of the article.

▶ Write three supporting details.

▶ Read what you have written. Cross out anything that does not directly support the main idea.

▶ Exchange papers with your partner and ask him or her to check your choice of a main idea. Have your partner also check your choice of supporting details.

Summarize Write your summary on a separate piece of paper. Use your own words. Include the main ideas and details you wrote.

© Macmillan/McGraw-Hill

Electricity

Use your textbook to help you fill in the blanks.

What is electrical charge?

1. Volume, mass, and _____electrical charge_____ are properties of matter.

2. The two types of electrical charge are _____positive_____ charge and _____negative_____ charge.

3. Objects with opposite charges _____attract_____ each other.

4. Objects with the same charge _____repel_____ each other.

5. Electricity is made up of _____charged particles_____ .

6. Charge is _____balanced_____ on most objects, meaning that they have equal numbers of positive and negative particles.

7. The buildup of electrical charge on an object is _____static electricity_____ .

What is electric current?

8. The flow of an electric charge from one place to another is called _____electric current_____ .

9. The energy from electric currents is used to produce

_____heat_____ , light, and _____motion_____ .

10. The path on which current flows is a(n) _____circuit_____ .

11. The flow of current can be controlled by a(n) _____switch_____ .

12. When a switch is turned off, a _____circuit_____ is open
and current does not flow.

13. When a switch is on, the circuit is _____closed_____

and current _____flows_____ .

What are conductors and insulators?

14. Current flows easily through _____conductors_____ ,
which are made of metals such as copper.

15. Current does not flow easily through _____insulators_____ ,
which are made of materials such as plastic.

Critical Thinking

16. Why do electricians wrap copper wires with black
plastic tape?

Electricians wrap copper wires with black plastic tape to

insulate the wires so that they cannot pass electricity to things

that touch them.

_____

_____

© Macmillan/McGraw-Hill

Electricity

What am I?

Choose a word from the box below that answers each question. Then write its letter in the space provided.

a. battery	**d.** conductor	**g.** insulator
b. circuit	**e.** electrical charge	**h.** negative charge
c. closed circuit	**f.** electric current	

1. _____d_____ I allow current to flow easily. What am I?

2. _____b_____ I am the path for electric current. What am I?

3. _____e_____ I can be positive or negative. What am I?

4. _____g_____ I keep charges from flowing easily. What am I?

5. _____a_____ I am a power source that provides an electric charge. What am I?

6. _____f_____ I am a flow of electric charge. What am I?

7. _____h_____ I repel negative charges. What am I?

8. _____c_____ My switch is off. What am I?

Electricity

Use the words in the box to fill in the blanks below.

balanced	closed	electric current	positive
circuit	conductors	open	switch

People depend on electricity for many things.

Electricity is made of particles of matter that have

_____positive_____ or negative electrical charges. Most

objects have an equal number of positive and negative

charges, or a(n) _____balanced_____ charge.

Charge that is flowing is a(n) ___electric current___ .

An electric current needs a path, or _____circuit_____ ,

through which to flow. Current flows easily through

_____conductors_____ , such as copper, and poorly through

insulators, such as glass and plastic. A circuit may have

a(n) _____switch_____ that opens or closes the circuit.

Current will flow when the switch is _____closed_____ ,

but not when the switch is _____open_____ .

Electrical circuits are an important part of many

objects that you use every day.

© Macmillan/McGraw-Hill

Other Energy Sources

Write About It

Persuasive Writing Write a persuasive letter to a community leader. Tell why you think it is important to find other sources of energy. Be sure to follow the form of a formal letter.

Getting Ideas

Use the chart below to help you get started. Write your opinion in the top box. Write convincing reasons, facts, and examples in the bottom boxes.

Opinion: We must find other energy sources before it's too late.

Oil, coal, and gas cannot be replaced easily.	We hurt the land when we look for oil, coal, and gas.	We depend on other nations for oil.

Planning and Organizing

Here are three sentences that Daria wrote. Does the sentence support the opinion that we need to find other sources of energy? If so, write "yes." If not, write "no."

1. _____yes_____ We hurt the land when we mine for coal.

2. ___yes___ We are using up our supply of oil.

3. ___no___ Some cars run on electricity as well as oil.

Drafting

Write a sentence to begin your letter. Tell your opinion about finding other sources of energy.

Possible sentence: We must find other sources of energy before it's

too late.

Now write your letter. Use a separate piece of paper. Follow the format of a formal letter. Begin with the sentence above. Then give facts, reasons, and examples to support it.

Revising and Proofreading

Here is a part of Daria's letter. She made five punctuation errors. Find the mistakes and correct them.

> Dear Mr Alvarez
>
>   We must find other sources of energy before its too late.
> We are using up our oil? Will there be any left when I am
> grown up. We are hurting the land by digging for coal.

Now revise and proofread your writing. Ask yourself:

▶ Did I follow the format of a formal letter?

▶ Did I clearly tell my opinion?

▶ Did I include convincing facts, reasons, and examples?

▶ Did I correct all mistakes?

© Macmillan/McGraw-Hill

Forms of Energy

Circle the letter of the best answer.

1. Which of the following materials would make a good conductor?

   a. glass

   b. copper

   c. plastic

   d. rubber

2. A sound is produced when something

   a. contracts.

   b. expands.

   c. gets smaller.

   d. vibrates.

3. You see a red flower when

   a. only the red color in white light is absorbed.

   b. every color but red is reflected.

   c. every color but white is absorbed.

   d. only the red color in white light is reflected.

4. When a thermometer gains thermal energy, the particles of liquid in the thermometer

   a. slow down.

   b. expand.

   c. contract.

   d. become cooler.

5. Earth's main source of heat is

   a. batteries.

   b. natural gas.

   c. the Sun.

   d. hot water.

6. The pitch of a sound is how

   a. high or low the sound is.

   b. warm or cool the sound is.

   c. close to the ear the sound is.

   d. loud the sound is.

© Macmillan/McGraw-Hill

Circle the letter of the best answer.

7. Light reflected from an object

   a. bends the object.

   b. bounces off the object.

   c. makes a dark area called a shadow.

   d. passes through the object.

8. A tool that separates white light into its different colors is a

   a. conductor.

   b. prism.

   c. switch.

   d. thermometer.

9. When warm and cool objects touch, what moves from the warmer object to the cooler object?

   a. charged particles

   b. sound waves

   c. heat

   d. electric current

10. What describes how much thermal energy an object has?

   a. mass

   b. volume

   c. color

   d. temperature

11. Sound travels best through

   a. solids.

   b. gases.

   c. liquids.

   d. space.

12. What part of your eye sends messages from light to the brain?

   a. cornea

   b. optic nerve

   c. pupil

   d. lens

© Macmillan/McGraw-Hill